1983

PERGAMON INTERNATIONA
of Science, Technology, Engineering and

The 1000-volume original paperback library i
industrial training and the enjoyment of leisure

Publisher: Robert Maxwell, M.C.

Extraterrestrials
WHERE
ARE
THEY
?

Pergamon Titles of Related Interest

Billingham/Pesek COMMUNICATION WITH EXTRATERRESTRIAL INTELLIGENCE
Curtin TRENDS IN COMMUNICATIONS SATELLITES

Related Journals*

CHINESE ASTRONOMY
ADVANCES IN SPACE EXPLORATION
FUTURICS
SPACE RESEARCH
VISTAS IN ASTRONOMY

*Free specimen copies available upon request.

Extraterrestrials
WHERE ARE THEY ?

Edited by Michael H. Hart and Ben Zuckerman

PERGAMON PRESS

New York   Oxford   Toronto   Sydney   Paris   Frankfurt

Pergamon Press Offices:

U.S.A.   Pergamon Press Inc., Maxwell House, Fairview Park,
                         Elmsford, New York 10523, U.S.A.

U.K.   Pergamon Press Ltd., Headington Hill Hall,
                         Oxford OX3 OBW, England

CANADA   Pergamon Press Canada Ltd., Suite 104, 150 Consumers Road,
                         Willowdale, Ontario M2J 1P9, Canada

AUSTRALIA   Pergamon Press (Aust.) Pty. Ltd., P.O. Box 544,
                         Potts Point, NSW 2011, Australia

FRANCE   Pergamon Press SARL, 24 rue des Ecoles,
                         75240 Paris, Cedex 05, France

FEDERAL REPUBLIC   Pergamon Press GmbH, Hammerweg 6,
OF GERMANY   6242 Kronberg/Taunus, Federal Republic of Germany

Library of Congress Cataloging in Publication Data
Main entry under title:

Extraterrestrials-where are they?

   Bibliography: p.
   Includes index.
   1. Life on other planets--Congresses. 2. Life--
Origin--Congresses. I. Hart, Michael H.
II. Zuckerman, Ben, 1943-
QB54.E95   1981   574.999   81-1876
ISBN 0-08-026342-9   AACR2
ISBN 0-08-026341-0 (pbk.)

Printed in the United States of America

Contents

v

Preface

Where Are They? Enrico Fermi is reputed to have asked this question at the dawn of the atomic age. He must have been wondering why, having discovered and tamed nuclear energy sources, advanced extraterrestrials were not in evidence here on Earth or out in the skies.

During the 1960's and early 1970's, Fermi's question was largely forgotten or ignored. Advances in radio astronomy, the American and Soviet space programs, the blossoming of the study of molecular biology, and progress in laboratory simulations of prebiological organic chemistry all contributed, in their own way, to a euphoric belief among many scientists that life in the cosmos is commonplace and might even be discovered soon. At a more popular level, numerous reports of close encounters of the second and third kind, lavishly bankrolled science fiction movies, and enormously popular books on ancient astronauts all served to promote the idea that They are out there and will soon be, or already have been, here.

The past few years have seen the introduction of new and sobering input into this picture. The U.S. program of planetary exploration, while highly successful from a technological and scientific standpoint, has failed to produce even a hint of an extraterrestrial biology. Although the search for simple non-terrestrial life in our solar system cannot be considered complete, the prospects for eventual success do not look good. In addition, searches for evidence of advanced technology, either in deep space or in the solar system, have been discouraging. To astronomers who work with giant optical and radio telescopes the Universe appears to be a gigantic wilderness area untouched by the hand of intelligence (with the possible exception of God's). The absence of advanced extraterrestrials on the Earth and, very probably, in the solar system has been interpreted by various scientists, beginning with Michael Hart's 1975 paper that leads off this volume, as evidence that such creatures may not exist anywhere in our Milky Way galaxy. Thus, Fermi's question has now reemerged to haunt our minds.

Sometime in 1978 it occurred to Drs. Hart, Papagiannis and myself that the scientific community was scarcely aware of many of these ideas and the public at large even less so. So we organized a modest meeting – "Where Are They? A Symposium on the Implications of Our Failure to Observe Extraterrestrials" – that was held in College Park, Maryland, at the Center of Adult Education of the University of Maryland on November 2 and 3, 1979. The size of the symposium was kept small to encourage and facilitate back and forth discussion among the participants. Some of this discussion, which was tape recorded, is included in the present book following the various papers.

The first day of the conference was devoted to a general overview of the question and the current observational situation, followed by a session on the feasibility of interstellar travel and colonization. On the second day the biological and astronomical contexts of the problem were explored. We believe that many of the papers that were presented at the symposium and in the present volume suggest a significantly different picture from the one that has been painted in most earlier meetings, NASA supported studies, and compendia on the search for extraterrestrial intelligence.

An image of the Milky Way abounding in all kinds of advanced super-civilizations and bizarre forms of life is tremendously appealing to many scientists and lay persons. Then life is but a commonplace extension of cosmic evolution following the Big Bang and we human beings are insignificant – mere cosmic insects. Perhaps this accounts in small part for the indifferent and even ruthless way that many human beings treat life on Earth.

But what if this is not so? What if after 10 billion years and hundreds of billions of stars we possess the most advanced brains in the galaxy? Surely, if we are alone, then we were meant to play a more noble role than that currently in evidence on our troubled globe. If we are to travel to the stars in the next millennium, then we must, in the next century at most, gain control of our runaway population, rapacious appetites, and nuclear and biological technologies. If we fail to do this, then the radiation of intelligence throughout the Milky Way may have to wait for a distant time and place.

Ben Zuckerman
College Park, Maryland
December, 1980

Acknowledgments

We gratefully acknowledge the assistance of Dr. John Trasco and Ms. Margaret Berry, and many other members of the Astronomy Program at the University of Maryland without which our meeting could not have taken place. The Astronomy Program and the Division of Mathematical and Physical Sciences and Engineering very kindly supplied financial aid for the meeting. Mr. Elliot Eichler took the group photo that appears in this book. Most of all we wish to thank Dr. Michael Papagiannis for his help in planning and organizing the symposium.

Acknowledgments

1

An Explanation for the Absence of Extraterrestrials on Earth*
Michael H. Hart

ABSTRACT

We observe that no intelligent beings from outer space are now present on Earth. It is suggested that this fact can best be explained by the hypothesis that there are no other advanced civilizations in our Galaxy. Reasons are given for rejecting all alternative explanations of the absence of extraterrestrials from Earth.

Are there intelligent beings elsewhere in our Galaxy? This is the question which astronomers are most frequently asked by laymen. The question is not a foolish one; indeed, it is perhaps the most significant of all questions in astronomy. In investigating the problem we must therefore do our best to include all relevant observational data.

Because of our training, most scientists have a tendency to disregard all information which is not the result of measurements. This is, in most matters, a sensible precaution against the intrusion of metaphysical arguments. In the present matter, however, that policy has caused many of us to disregard a clearly empirical fact of great importance, to wit: There are no intelligent beings from outer space on Earth now. (There may have been visitors in the past, but none of them has remained to settle or colonize here.) Since frequent reference will be made to the foregoing piece of data, in what follows we shall refer to it as 'Fact A'.

Fact A, like all facts, requires an explanation. Once this is recognized, an argument is suggested which indicates an answer to our original question. If, the argument goes, there were intelligent beings elsewhere in our Galaxy, then they would eventually have achieved space travel, and would have explored and colonized the Galaxy, as we

*Reprinted from the Quarterly Journal of the Royal Astronomical Society, 16, 128-35, (1975), with permission of Blackwell Scientific Publications, Ltd.

have explored and colonized the Earth. However, (Fact A), they are not here; therefore they do not exist.

The author believes that the above argument is basically correct; however, in the rather loose form stated above it is clearly incomplete. After all, might there not be some other explanation of Fact A? Indeed, many other explanations of Fact A have been proposed: however, none of them appears to be adequate.

The other proposed explanations of Fact A might be grouped as follows:

(1) All explanations which claim that extraterrestrial visitors have never arrived on earth because some physical, astronomical, biological or engineering difficulty makes space travel infeasible. We shall refer to these as 'physical explanations'.

(2) Explanations based on the view that extraterrestrials have not arrived on Earth because they have chosen not to. This category is also intended to include any explanation based on their supposed lack of interest, motivation or organization, as well as political explanations. We shall refer to these as 'sociological explanations'.

(3) Explanations based on the possibility that advanced civilizations have arisen so recently that, although capable and willing to visit us, they have not had time to reach us yet. We shall call these 'temporal explanations'.

(4) Those explanations which take the view that the Earth has been visited by extraterrestrials, though we do not observe them here at present.

These four categories are intended to be exhaustive of the plausible alternatives to the explanation we suggest. Therefore, if the reasoning in the next four sections should prove persuasive, it would seem very likely that we are the only intelligent beings in our Galaxy.

PHYSICAL EXPLANATIONS

After the success of Apollo 11 it seems strange to hear people claim that space travel is impossible. Still, the problems involved in interstellar travel are admittedly greater than those involved in a trip to the Moon, so it is reasonable to consider just how serious the problems are, and how they might be overcome.

The most obvious obstacle to interstellar travel is the enormity of the distances between the stars, and the consequently large travel times involved. A brief computation should make the difficulty clear: The greatest speeds which manned aircraft, or even spacecraft, have yet attained is only a few thousand km hr^{-1}. Yet even travelling at 10 per cent of the speed of light ($\sim$one billion km hr^{-1}) a one-way trip to Sirius, which is one of the nearest stars, would take 88 years. Plainly, the problem presented is not trivial; however, there are several possible means of dealing with it:

(1) If it is considered essential that those who start on the voyage should still be reasonably youthful upon arrival, this could be accomplished by having the voyagers spend most of the trip in some form of 'suspended animation'. For example, a suitable combination of drugs might not only put a traveller to sleep, but also slow his metabolism down by a factor of 100 or more. The same result might be effected by freezing the space voyagers near the beginning of the trip, and thawing them out shortly before arrival. It is true that we do not yet know how to freeze and revive warm-blooded animals but: (a) future biologists on Earth (or biologists in advanced civilizations elsewhere) may learn how to do so; (b) intelligent beings arising in other solar systems are not necessarily warm-blooded.

(2) There is no reason to assume that all intelligent extraterrestrials have life spans similar to ours. (In fact, future medical advances may result in human beings having life expectancies of several millennia, or even perhaps much longer.) For a being with a life span of 3000 years a voyage of 200 years might seem not a dreary waste of most of one's life, but rather a diverting interlude.

(3) Various highly speculative methods of overcoming the problem have been proposed. For example, utilization of the relativistic time-dilation effect has been suggested (though the difficulties in this approach seem extremely great to me). Or the spaceship might be 'manned' by robots, perhaps with a supplementary population of frozen zygotes which, after arrival at the destination, could be thawed out and used to produce a population of living beings.

(4) The most direct manner of handling the problem, and the one which makes the fewest demands on future scientific advances, is the straightforward one of planning each space voyage, from the beginning, as one that will take more than one generation to complete. If the spaceship is large and comfortable, and the social structure and arrangements are planned carefully, there is no reason why this need be impracticable.

Another frequently mentioned obstacle to interstellar travel is the magnitude of the energy requirements. This problem might be insurmountable if only chemical fuels were available, but if nuclear energy is used the fuel requirements do not appear to be extreme. For example, the kinetic energy of a spaceship travelling at one-tenth the speed of light is:

$$KE = (\gamma - 1) Mc^2 = ((1.0 - 0.01)^{-1/2} - 1) Mc^2 = 0.005 \ Mc^2 \qquad (1)$$

Now the energy released in the fusion of a mass F of hydrogen into helium is approximately 0.007 Fc^2. In principle, the mechanical efficiency of a nuclear-powered rocket can be more than 60 percent (1, 2). However, let us assume that in practice only one-third of the nuclear energy could actually be released and converted into kinetic energy of the spacecraft. Then the fuel needed to accelerate the spaceship to 0.10 c is given by:

$$0.005 \ Mc^2 = 0.007 \ Fc^2/3. \qquad\qquad (2)$$

This gives: F = 2.14 M, and T = 3.14 M, where T is the combined mass of spaceship and fuel. The necessity of starting out with enough fuel first to accelerate the ship, and later to decelerate, introduces another factor of 3.14; so initially we must have T = 9.88 M. In other words, the ship must starts its voyage carrying about nine times its own weight in fuel. This is a rather modest requirement, particularly in view of the cheapness and abundance of the fuel. (The enormous fuel-to-payload ratios computed by Purcell (3) are a result of his considering only relativistic space flight; a travel speed of 0.1 c seems more realistic.) Furthermore, there are several possible ways of reducing the fuel-to-payload ratio, including (a) refuelling from auxiliary craft; (b) scooping up H atoms while travelling through interstellar space; (c) greater engine efficiencies; (d) travelling at slightly lower speeds (travelling at 0.09 c instead of 0.10 c would reduce the fuel-to-payload ratio to 6.5:1); and (e) using methods of propulsion other than rockets. For some interesting possibilities see Marx(4) and other papers listed by Mallove & Forward(5).

It can be seen that neither the time of travel nor the energy requirements create an insuperable obstacle to space travel. However, in the past, it was sometimes suggested that one or more of the following would make space travel unreasonably hazardous: (a) the effects of cosmic rays; (b) the danger of collisions with meteoroids; (c) the biological effects of prolonged weightlessness; and (d) unpredictable or unspecified dangers. With the success of the Apollo and Skylab missions it appears that none of these hazards is so great as to prohibit space travel.

SOCIOLOGICAL EXPLANATIONS

Most proposed explanations of Fact A fall into this category. A few typical examples are:

(a) Why take the anthropomorphic view that extraterrestrials are just like us? Perhaps most advanced civilizations are primarily concerned with spiritual contemplation and have no interest in space exploration. (The Contemplation Hypothesis.)

(b) Perhaps most technologically advanced species destroy themselves in nuclear warfare not long after they discover atomic energy. (The Self-Destruction Hypothesis.)

(c) Perhaps an advanced civilization has set the Earth aside as their version of a national forest, or wildlife preserve. (The Zoo Hypothesis(6).)

In addition to variations on these themes (for example, extraterrestrials might be primarily concerned with artistic values rather than spiritual contemplation) many quite different explanations have been suggested. Plainly, it is not possible to consider each of these individually. There is, however, a weak spot which is common to all of these theories.

Consider, for example, the Contemplation Hypothesis. This might be a perfectly adequate explanation of why, in the year 600,000 BC, the inhabitants of Vega III chose not to visit the Earth. However, as we well know, civilizations and cultures change. The Vegans of 599,000 BC could well be less interested in spiritual matters than their ancestors were, and more interested in space travel. A similar possibility would exist in 598,000 BC, and so forth. Even if we assume that the Vegans' social and political structure is so rigid that no changes occur even over hundreds of thousands of years, or that their basic psychological makeup is such that they always remain uninterested in space travel, there is still a problem. With such an additional assumption the Contemplation Hypothesis might explain why the Vegans have never visited the Earth, but it still would not explain why the civilizations which developed on Procyon VI, Sirius II, and Altair IV have also failed to come here. The Contemplation Hypothesis is not sufficient to explain Fact A unless we assume that it will hold for every race of extra-terrestrials — regardless of its biological, psychological, social or political structure — and at every stage in their history after they achieve the ability to engage in space travel. That assumption is not plausible, however, so the Contemplation Hypothesis must be rejected as insufficient.

The same objection, however, applies to any other proposed sociological explanation. No such hypothesis is sufficient to explain Fact A unless we can show that it will apply to every race in the Galaxy, and at every time.

The foregoing objection would hold even if there were some established sociological theory which predicted that most technologically advanced civilizations will be spiritually oriented, or will blow themselves up, or will refrain from exploring and colonizing. In point of fact, however, there is no such theory which has been generally accepted by political scientists, or sociologists, or psychologists. Furthermore, it is safe to say that no such theory will be accepted. For any scientific theory must be based upon evidence, and the only evidence concerning the behaviour of technologically advanced civilizations which political scientists, sociologists and psychologists have comes from the human species — a species which has neither blown itself up, nor confined itself exclusively to spiritual contemplation, but which has explored and colonized every portion of the globe it could. (This is not intended as proof that all extraterrestrials must behave as we have; it is intended to show that we cannot expect a scientific theory to be developed which predicts that most extraterrestrials will behave in the reverse way.)

Another objection to any sociological explanation of Fact A is methodological. Faced with a clear physical fact astronomers should attempt to find a scientific explanation for it — one based on known physical laws and subject to observational or experimental tests. No scientific procedure has ever been suggested for testing the validity of the Zoo Hypothesis, the Self-Destruction Hypothesis, or any other suggested sociological explanation of Fact A; therefore to accept any such explanation would be to abandon our scientific approach to the question.

TEMPORAL EXPLANATIONS

Even if one rejects the physical and sociological explanations of Fact A, the possibility exists that the reason no extraterrestrials are here is simply because none have yet had the time to reach us. To judge how plausible this explanation is, one needs some estimate of how long it might take a civilization to reach us once it had embarked upon a programme of space exploration. To obtain such an estimate, let us reverse the question and ask how long it will be, assuming that we are indeed the first species in our Galaxy to achieve interstellar travel, before we visit a given planet in the Galaxy?

Assume that we eventually send expeditions to each of the 100 nearest stars. (These are all within 20 light-years of the Sun.) Each of these colonies has the potential of eventually sending out their own expeditions, and their colonies in turn can colonize, and so forth. If there were no pause between trips, the frontier of space exploration would then lie roughly on the surface of a sphere whose radius was increasing at a speed of 0.10 c. At that rate, most of our Galaxy would be traversed within 650,000 years. If we assume that the time between voyages is of the same order as the length of a single voyage, then the time needed to span the Galaxy will be roughly doubled.

We see that if there were other advanced civilizations in our Galaxy they would have had ample time to reach us, unless they commenced space exploration less than 2 million years ago. (There is no real chance of the Sun being accidentally overlooked. Even if the residents of one nearby planetary system ignored us, within a few thousand years an expedition from one of their colonies, or from some other nearby planetary system, would visit the solar system.)

Now the age of our Galaxy is ~ 10^{10} years. To accept the temporal explanation of Fact A we must therefore hypothesize that (a) it took roughly 5000 time-units (choosing one time-unit $\equiv 2 \times 10^6$ years) for the first species to arise in our Galaxy which had the inclination and ability to engage in interstellar travel; but (b) the second such species (i.e. us) arose less than 1 time-unit later.

Plainly, this would involve a quite remarkable coincidence. We conclude that, though the temporal explanation is theoretically possible, it should be considered highly unlikely.

PERHAPS THEY HAVE COME

There are several versions of this theory. Perhaps the most common one is the hypothesis that visitors from space arrived here in the fairly recent past (within, say, the last 5000 years) but did not settle here permanently. There are various interesting archaeological finds which proponents of this hypothesis often suggest are relics of the aliens' visit to Earth.

The weak spot of that hypothesis is that it fails to explain why the Earth was not visited earlier:

(a) If it is assumed that extraterrestrials have been able to visit us for a long time, then a sociological theory is required to explain why they all postponed the voyage to Earth for so long. However, any such sociological explanation runs into the same difficulties described earlier.

(b) On the other hand, suppose it is assumed that extraterrestrials visited us as soon as they were able to. That this occurred within 5000 years (which is only 1/400 of a time-unit) of the advent of our own space age would involve an even more remarkable coincidence than that discussed in the previous section.

Another version of the theory is that the Earth was visited from space a very long time ago, say 50 million years ago. This version involves no temporal coincidence. However, once again, a sociological theory is required to explain why, in all the intervening years, no other extraterrestrials have chosen to come to Earth, and remain. Of course, any suggested mechanism which is effective only 50 percent (or even 90 percent) of the time would be insufficient to explain Fact A. (For example, the hypothesis that most extraterrestrials wished only to visit, but not to colonize, is inadequate. For colonization not to have occurred requires that every single civilization which had the opportunity to colonize chose not to.)

A third version, which we may call 'the UFO Hypothesis', is that extraterrestrials have not only arrived on Earth, but are still here. This version is not really an explanation of Fact A, but rather a denial of it. Since very few astronomers believe the UFO Hypothesis it seems unnecessary to discuss my own reasons for rejecting it.

CONCLUSIONS AND DISCUSSION

In recent years several astronomers have suggested that intelligent life in our Galaxy is very common. It has been argued(7) that (a) a high percentage of stars have planetary systems; (b) most of these systems contain an Earth-like planet; (c) life has developed on most of such planets; and (d) intelligent life has evolved on a considerable number of such planets. These optimistic conclusions have perhaps led many persons to believe that (1) our starfaring descendants are almost certain, sooner or later, to encounter other advanced cultures in our Galaxy; and (2) radio contact with other civilizations may be just around the corner.

These are very exciting prospects indeed; so much so that wishful thinking may lead us to overestimate the chances that the conjecture is correct. Unfortunately, though, the idea that thousands of advanced civilizations are scattered throughout the Galaxy is quite implausible in the light of Fact A. Though it is possible that one or two civilizations have evolved and have destroyed themselves in a nuclear war, it is implausible that every one of 10 000 alien civilizations had done so. Our descendants might eventually encounter a few advanced civilizations which never chose to engage in interstellar travel; but their number should be small, and could well be zero.

If the basic thesis of this paper is correct there are two corollary conclusions: (1) an extensive search for radio messages from other civilizations is probably a waste of time and money; and (2) in the long run, cultures descended directly from ours will probably occupy most of the habitable planets in our Galaxy.

In view of the enormous number of stars in our Galaxy, the conclusions reached in this paper may be rather surprising. It is natural to inquire how it has come about that intelligent life has evolved on Earth in advance of its appearance on other planets. Future research in such fields as biochemistry; the dynamics of planetary formation; and the formation and evolution of atmospheres, may well provide a convincing answer to this question. In the meantime, Fact A provides strong evidence that we are the first civilization in our Galaxy, even though the cause of our priority is not yet known.

REFERENCES

(1) Von Hoerner, S. (1962). Science, 137, 18-23.

(2) Marx, G. (1963). Astr. Acta, 9, 131-139.

(3) Purcell, E. (1963). In Interstellar Communication. (editor: A.G.W. Cameron), Benjamin, New York. Chapter 13.

(4) Marx, G. (1966). Nature, 211, 22-23.

(5) Mallove, E.F. & Forward, R.L. (1972). Bibliography of interstellar travel and communication, Research Report 460, pp. 16-21, Hughes Research Laboratories, Malibu, California.

(6) Ball, John A. (1973). Icarus, 19, 347-349.

(7) Shklovskii, I.S. & Sagan, C. (1968). Intelligent life in the Universe, Chapter 29, Holden-Day, San Francisco.

2

Searches for Electromagnetic Signals from Extraterrestrial Beings

Ben Zuckerman

ABSTRACT

There have been more than a dozen searches for radio signals from alien civilizations. Targets have included nearby stars and external galaxies. We briefly review a few of the most ambitious and sensitive projects. Even if the chance of success is small, it seems inevitable that such endeavors will be continued into the foreseeable future since their cost is also small and the potential thrills to the experimenter may be very large.

Within the next few years infrared telescopes in orbit around the Earth should greatly enhance our ability to detect very advanced civilizations (Dyson spheres). We briefly consider the relative merits of infrared and radio wavelengths for interstellar communication between two civilizations that possess advanced space technologies.

I. INTRODUCTION

Discovery of and communication with an extraterrestrial civilization by means of electromagnetic radiation requires a cooperative effort. Success can be achieved only if time and money are expended at both the transmitting and receiving ends. This contrasts with the passive mode in which they do all the work (by sending their flying saucers to our solar system) or the active mode in which we do all the work (by sending probes to other stars).

The radiation that passes between two different star systems may be generated specifically for the purpose of interstellar communication or it may be produced for other reasons but with sufficient intensity to be detected over distances of ten or more light years. To date almost all attempts to detect interstellar signals have been carried out at radio wavelengths. Is this only an historical accident due to chance developments in Western technology combined, perhaps, with the high transparency of the terrestrial atmosphere at radio wavelengths? Many groups have argued that there are fundamental reasons why all civilizations will use radio waves for interstellar communication (e.g., Oliver and Billingham 1971; Billingham et al. 1979). But this opinion is not unanimous and Townes (1981) argues that, for many purposes, the infrared may eventually be superior to the microwave domain.

This debate cannot be settled at the present time because civilizations only slightly in advance of our own are likely to have constructed very large telescopes in space. The relative merits of the infrared and radio domains depend sensitively on factors such as $A_T(\lambda)$ and $A_R(\lambda)$, the collecting areas of the transmitting and receiving antennas. Our current knowledge of space technology is still insufficient to predict either $A_T(\lambda)$ or $A_R(\lambda)$, since λ, the wavelength of the interstellar signal, may be as long as 21-cm or short as 10 µm (a difference of a factor of more than 10,000). Additional considerations, for example scattering in interstellar and interplanetary space, may also bias the ultimate choice of wavelength (see caption to Figure 2.1).

II. SEARCHES AT RADIO WAVELENGTHS

Radio signals from extraterrestrial intelligence (ETI) have been searched for by astronomers in the USA, USSR and Canada. The total number of such programs has been fewer than two dozen, most of which were of significantly lower sensitivity than the six that are listed in Table 2.1. In the West, little is known about recent Soviet investigations (Billingham et al. 1979). Unless someone is hiding something in their bottom drawer, all searches have yielded negative results.

We will briefly discuss the six projects in Table 2.1. Five of these were carried out in the USA and one in Canada. Patrick Palmer and Ben Zuckerman (hereafter PZ) were the first to exploit the very considerable advances in instrumentation in spectral line radio astronomy in the 1960's. They pointed the 91-meter transit telescope of the U.S. National Radio Astronomy Observatory toward stars located within 80 light years of the Earth that are listed in the Gliese (1969) and RGO (Wooley et al. 1970) catalogs. PZ limited their search to F, G, K, and M main sequence stars since, as has been argued many times in the literature, these are the stars that are most likely to provide a habitable zone that is stable for at least a billion years. This may be about the minimum time necessary for life to originate and evolve a technological civilization. These limits in spectral type may be much too broad if the continuously habitable zone is as narrow as suggested by the calculations of Hart (1979) and creatures never leave their own planetary system. On the other hand, if, thanks to interstellar travel, the galaxy has already been physically settled, then many (most) nearby star systems should harbor technological civilizations. In this case, essentially all main-sequence stars and possibly red giant stars as well should be examined for radio signals (e.g., Kuiper and Morris 1977). Thus, although the programs listed in Table 2.1 are quite modest in many respects, they already suggest that our Milky Way galaxy has not been extensively colonized.

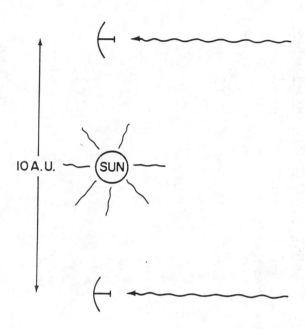

Fig. 2.1. A giant space interferometer (after Buyakis et al. 1979). The telescopes are 10 km in diameter and they are separated by 10 astronomical units (i.e., ten times the distance between the Earth and the Sun). The angular resolution of such a system at 1 millimeter wavelength (where the entire Universe is in the near field) is $\sim 10^{-10}$ second of arc. In practice the angular resolution may be limited by scattering from plasma irregularities in interstellar and interplanetary space. Since this effect varies as λ^2, better angular resolution should be obtainable in the infrared. At even greater distances from the Sun, it should be possible to construct even larger telescopes. (Note that in the figure the telescopes and the Sun are not to scale.)

Most stars are members of multiple star systems. PZ examined numerous binary stars in which the star-star separation is either less than about one-third or greater than about three times the radius of the habitable zone around the primary star. According to Harrington (1977), the orbits of planets in this zone will be stable for billions of years. However, we still do not know if planets are able to form in systems that contain more than one star.

TABLE 2.1. SENSITIVE RADIO SEARCHES FOR ETI

DATE	TELESCOPE	OBSERVER	TARGET	WAVELENGTH	SPECTRAL RESOLUTION	SENSITIVITY (W/m^2)
1972-76	91 m (NRAO)	Palmer Zuckerman	670 nearby stars	21 cm	4 kHz	10^{-23}
1973-present	53 m (Ohio State)	Dixon Cole	all sky	21 cm	10 kHz	few x 10^{-21}
1974-limbo	46 m (Algonquin Park)	Feldman Bridle	70 nearby stars	1.3 cm	30 kHz	few x 10^{-22}
1975-76	305 m (Arecibo)	Drake Sagan	several nearby galaxies	12.5 cm 18 cm 21 cm	1 kHz	10^{-24}
1977	91 m (NRAO)	Tarter Cuzzi Black Clark	200 nearby stars	18 cm	5.5 Hz	few x 10^{-24}
1978	305 m (Arecibo)	Horowitz	185 nearby stars	21 cm	0.015 Hz	few x 10^{-27}

All of the programs listed in Table 2.1 involved looking for signals of narrow bandwidth. If, as seems plausible, the total power available for transmission from an extraterrestrial civilization is independent of the transmitter bandwidth, then the signal-to-noise ratio at the Earth can be maximized when this power is concentrated into the narrowest possible band of frequencies. In addition, very narrow signals can easily be distinguished from emission from interstellar atoms and molecules which is not likely to be narrower than either a thermal bandwidth corresponding to temperatures ~ 10 K or a line-narrowed unsaturated maser of comparable width.

In their search, PZ used the 384-channel autocorrelation receiver of the U.S. National Radio Astronomy Observatory. Half of the channels covered a total bandwidth of 10 MHz and the other half covered 625 KHz. The latter 192 channels were each 4 KHz wide, just narrow enough to discriminate against emission by interstellar hydrogen atoms. The passband of the autocorrelator was centered at the 21-cm wavelength of the "spin-flip" (hyperfine) transition of atomic hydrogen in the rest frame of each star. A typical star was observed for seven days in succession for about four minutes per day. One of the signposts for signals from ETI is rapid variability (timescales of seconds to days). PZ observed "glitches" – time variable signals – in about 10 stars. These stars were then reobserved, usually after a delay of about one year. In only one such case was a second glitch observed. In no case were glitches present a sufficient fraction of the total observing time to justify much optimism that an alien civilization had actually been discovered. The major protection against terrestrial interference is an international agreement that forbids man-made transmissions in the 21-cm band. But occasional infringements do occur and could cause an occasional glitch.

Four of the six programs in Table 2.1 are targeted searches of nearby stars. The motivation for this strategy is as follows. The amount of power that civilizations that are not much in advance of our own (these are often called Kardashev Type I civilizations) are capable of broadcasting in all directions at any given time is such that, given existing radio telescopes here on Earth, we could not detect such radiation even from the nearest stars. So the searches listed in Table 2.1 are incapable, by orders of magnitude, of detecting leakage signals such as the TV carriers for Cosmos and Charlie's Angels if the extraterrestrials transmit at the same power levels as we now do (Sullivan et al. 1978). Therefore, if we are to detect a Type I civilization we must look for powerful transmitters that are beamed at the Earth. It seems that such beacons are not likely to be pointed in our direction except by our closest neighbors in space – hence the motivation for a search of nearby stars. PZ, for example, would have been able to detect a narrow band signal from a 40 megawatt transmitter on a 100-meter antenna located as far away as 80 light years.

Some of the very nearest stars examined in the four stellar search programs in Table 2.1 are close enough that an advanced technological civilization located near one of them could have detected leakage radiation from radio stations here on the Earth that were broadcasting during the 1930's and beamed back a reply that could have been received here during the 1970's. τ Ceti and ε Eridani, two sun-like stars that are located only about 10 light years from the Earth, satisfy this criterion. In 1960 Frank Drake, in his project Ozma, examined these two stars at length. PZ and others have also looked at these stars, more briefly than Drake, but with much more sensitive systems than were available in 1960.

Alan Bridle and Paul Feldman are carrying out a search program at a frequency near 22.2 GHz at the Canadian National Radio Observatory located in Algonquin Provincial Park. This frequency was chosen because it is one at which water vapor molecules can emit and absorb radiation. Of course, water is crucial for life on Earth, and it may be essential for all other forms of life in the Universe as well. In addition, H_2O molecules located near some stars and protostars are known to emit powerful beams of maser radiation at this 22.2 GHz frequency.

Drs. Bridle and Feldman searched a total bandwidth of 10 MHz with 30 KHz resolution. So far they have spent a total of about 140 hours observing 70 stars located within 50 light years of the Earth. This project is presently in cold storage but will be thawed out if and when a better 22 GHz receiver becomes available at Algonquin Park.

The program of Paul Horowitz (1978) was quite different from the PZ search even though they both examined nearby stars at 21-cm wavelength. Horowitz used a very large telescope and ultrahigh spectral resolution, thus he was very sensitive to very narrow signals but only over a limited range of phase space. His total bandwidth was only 1 KHz (65,536 equally spaced frequency bins each 0.015 Hz wide!). The choice of 0.015 Hz was dictated by two considerations: the short term stability of the rubidium reference oscillator used at the Arecibo Observatory and the ultimate limit to narrowband interstellar transmissions — line broadening by multiple scatterings from fluctuations in the ionized component of the interstellar medium.

Horowitz assumed that the transmitting civilization has accurately measured the radial velocity of our Sun and/or the velocity of the barycenter (center of mass) of our solar system. The two differ by, at most, 60 Hz at 21-cm wavelength. Because he searched a total bandwidth of only 1 KHz, Horowitz's strategy was to assume that They carefully control their transmitted signal such that it arrives in our solar system at the hydrogen-line rest frequency in either the solar or barycentric system. We note in passing that the most accurate measurements of radial velocity of solar type stars that we are likely to achieve within the next few decades will be at the 10 m/s level which corresponds to about 50 Hz at 21-cm wavelength. This accuracy will probably be limited by bulk motions in the stellar atmospheres.

Horowitz spent a total of 80 hours observing F, G, and K main sequence stars listed in the RGO catalog. He excluded all known binaries. For signal bandwidths less than 0.015 Hz, a megawatt transmitter on an Arecibo-sized antenna could have been detected out to distances of 3000 light years. Many such distant stars were included in the telescope beam although the stars that Horowitz was most interested in were all within 80 light years. He showed that 65,000 channel narrow-band spectrometers can be constructed rather easily and the data can be presented in such a way, a 256 x 256 raster "picture", that people can examine them conveniently. In addition, when observing with very narrow bandwidths terrestrial interference becomes a negligible problem. Horowitz had no glitches (false alarms) to worry about.

Jill Tarter et al. (1980) have also obtained excellent frequency resolution and many channels with yet a different innovative search technique. They recorded one-bit sampled data on high speed (720 kilobits/sec) Very Long Baseline Interferometer tape recorders. A magnetic tape is recorded in about 4 minutes which, quite by accident, is also about the maximum length of time that a star can be tracked with the NRAO 91-m transit telescope. The tapes, which are analyzed in post-real time, yield a total spectral coverage of 1.2 MHz at 5.5 Hz resolution – the equivalent of a 200,000 channel spectrum analyzer. Tarter et al. observed at the upper frequency end of the "water hole" (Oliver and Billingham 1971) near the frequencies of the interstellar hydroxyl (OH) transitions. Like Horowitz, they searched in the directions of (apparently) single main-sequence F, G, and K stars listed in the RGO catalog.

The Tarter et al. search is continuing and promises to yield the best frequency resolution until million-channel spectrum analyzers are constructed. Its disadvantages – modest instantaneous bandwidth and large computational overhead – can be partially compensated for with special purpose hardware processors and minicomputers.

The strategy employed in the four targeted search programs discussed above is to assume that technological civilizations exist near many stars in the Milky Way. If this is not the case (Hart 1975), then it may be necessary to search the distant reaches of the Milky Way or even another galaxy to find an ETI. A long lived technological civilization – one that exists for tens of thousands or even millions of years – may be capable of generating enormously powerful radio beacons that could be detected over immense distances. These advanced technologies are sometimes referred to as Kardeshev Type II or Type III civilizations (see Oliver and Billingham 1971, Appendix B).

An all-sky survey which covers most of the stars in our galaxy as well as in many other galaxies is currently underway at the Ohio State University Radio Observatory (Dixon 1979; Kraus 1979). The sky has been scanned, essentially continuously, since December 1973. By now about half of the sky has been examined, albeit for only a short time at each position. Although the sensitivity and spectral coverage (10 KHz resolution over a total bandwidth of 500 KHz) are quite modest, this program has the virtue of longevity and the possibility of extensive reobservation of positions that display interesting glitches. The most interesting signal detected during the first six years of searching is described by Kraus (1979).

Another program to search many distant stars was carried out at Arecibo Observatory by Frank Drake and Carl Sagan. Five nearby galaxies were observed for a total of about 100 hours. The hope was to find a super-civilization among the approximately 10^8 stars that were included within the telescope beam at each position. Leo I, Leo II, and M49 were each examined at nine positions, and M31 and M33 at many positions. The search covered 3 MHz of bandwidth at each of three separate wavelengths (21, 18, and 12.5 cm) at 1 KHz resolution. At the 21-cm (H) and 18-cm (OH) wavelengths the autocorrelation spectrometer was centered at the radial velocity of the known hydrogen emission at each position that was observed in M33. For the other galaxies, the spectrometer was centered at the systemic velocity if known or else at all possible systemic velocities of the local group.

III. THE FUTURE

Where do we go from here? At the least, modest microwave search programs similar to those described above are likely to continue into the foreseeable future. A substantial advance in sensitivity may be achieved during the 1980's by a joint venture of NASA's Ames Research Center and the Jet Propulsion Laboratory. They hope to obtain funding to construct narrow-band multimillion channel spectrometers and very low noise amplifiers to be coupled with the largest existing radio telescopes in the world. A ten year program is envisaged which would include both an all-sky survey and a targeted search of nearby and distant stars and star systems. If carried to completion this program could result in an improvement in detection capability for narrow-band signals from ETI by a factor of a million or more over the programs listed in Table 2.1 but at a cost of tens of millions of dollars. However, anticipated increases in utilization of the centimeter wavelength portion of the radio spectrum during the 1980's by various ground- and space-based services may severely compromise much of the NASA program.

A much more ambitious ground-based approach is to construct extensive arrays of large antennas such as the Cyclops project (Oliver and Billingham 1971). In addition to beamed transmissions from distant stars a Cyclops array could detect leakage radiation from nearby civilizations if they waste as much microwave power into space as we now do. Cyclops would cost many billions of dollars.

A different form of waste radiation might conceivably be detected in the near future by NASA's Infrared Astronomical Satellite scheduled for launch in late 1982. Advanced civilizations may reengineer their solar systems taking apart moons and even planets to construct habitats in space. If many such giant space arks are built, together they could intercept a substantial portion of the stellar radiation and reradiate it in the middle infrared (near 10 μ m). Such a system, sometimes referred to as a Dyson sphere, would produce an excess of infrared radiation in the otherwise normal spectrum of a main sequence star. IRAS could detect a modest Dyson sphere with an infrared luminosity equal to or greater than 1 percent of the luminosity of our Sun out to a distance of about 300 light years. This region contains some 10^5 stars. The advantage of searching for such radiation is that it is an unavoidable consequence of advanced astroengineering and does not imply the existence of a directional transmitter pointed at the Earth.

The steadily increasing level of man-made interference at radio wavelengths combined with the Earth's atmosphere, which is opaque at most infrared wavelengths, will help to push the search for electro-magnetic signals from ETI into interplanetary space. We mentioned in Section I that the relative merits of the infrared and microwave spectral regions for interstellar communication will depend ultimately on advanced space technologies. Buyakis et al. (1979) have considered some of the problems and potentials of constructing very large space telescopes, up to ten kilometers in diameter, in the outer reaches of the solar system (figure 2.1). Indeed it is not immediately obvious what sets the fundamental limitation to the size of telescopes that might be constructed out beyond the orbit of Neptune where gravitational deformations and the effects of differential solar heating are small. Conceivably, telescopes hundreds of kilometers in diameter may be constructed with the potential for detecting extremely weak signals from ETI.

A less speculative application of such telescopes would be spec-troscopic analysis of the atmospheric composition and conditions on planets orbiting about other stars. Then the existence of even non-technological forms of life may be established. For example, spec-troscopic detection of a strongly oxidizing atmosphere would have suggested the presence of life on Earth hundreds of millions of years before the emergence of Homo sapiens. At any rate, it should be much cheaper to study the universe with giant space telescopes than it would be to launch a myriad of probes to distant stars and galaxies. (The telescopes can also be used to control the probes should we choose to send them out [Zuckerman 1981].) In addition, we don't have to wait for a probe to reach its destination but rather we obtain an immediate payoff as soon as we point a giant telescope toward a distant star or galaxy. Giant space telescopes will tell us an immense amount about the universe and the prospects for life in it even if they fail to detect any signs of ETI.

This work was partially supported by National Science Foundation grant AST 76-17600 to the University of Maryland.

REFERENCES

Billingham, J., Pesek, R., and Seeger, C. (editors), (1979). Acta Astronautica 6 (numbers 1 and 2).

Buyakis, V.I., et al. (1979). Acta Astronautica 6, 175.

Dixon, R.S. (1979). Private communications.

Gliese, W., (1969). Catalogue of Nearby Stars, Veroffentlichungen Des Astron. Recheninstituts Heidelberg, #22.

Harrington, R.S. (1977). Astronomical Journal 82, 753.
Hart, M.H. (1975). Quarterly J. Roy. Astron. Soc. 16, 128.
Hart, M.H. (1979). Icarus 37, 351.
Horowitz, P. (1978). Science 201, 733.
Kraus, J. (1979). Cosmic Search 1, 31.
Kuiper, T.B.H. and Morris, M. (1977). Science 196, 616.
Oliver, B.M. and Billingham, J. (1971). Project Cyclops Report, NASA
    CR-114445.
Sullivan, W.T., Brown, S., Wetherill, C. (1978). Science 199, 377.
Tarter, J., Cuzzi, J., Black, D., and Clark T. (1980). Icarus 42, 136.
Townes, C.H. (1981). Proceedings of the National Academy of Sciences,
    in press.
Wooley, Sir R., Epps, E., Penston, M., and Pocock, S. (1970). Catalogue
    of Stars within 25 Parsecs of the Sun, Royal Observatory Annals
    (number 5). Herstmonceux, Royal Greenwich Observatory.
Zuckerman, B. (1981). J. British Interplanetary Society, in press.

DISCUSSION

Michael Hart: What was the cost of your program with Palmer, per star examined?

Zuckerman: We looked at about six hundred stars, at a total cost of about $30,000. So that comes out to only $50 per star.

Jill Tarter: Our cost per star was perhaps a factor of two more.

Gerald Feinberg: The cost per star must have been much smaller for the Drake/Sagan program.

Zuckerman: Yes, with 10^8 stars per position.

Feinberg: If each star was broadcasting at one times Arecibo, would that have been detectable?

Zuckerman: Yes.

Freeman Dyson: I think that the infrared observations could be done far better from the ground than from a satellite.

Zuckerman: IRAS will catalog everything that it sees in the infrared. In the scenario that I assumed, one percent of the stellar light is absorbed by the constructs of the advanced civilization and then reradiated in the infrared. The other 99 percent gets through as visible radiation. So you simply go out and cross check the IRAS catalog against the positions of known F, G, and K type stars. This is straightforward and relatively non-tedious. I'd be happy to check a million points myself, or at least get a graduate student to do it; and I think that the student would love to do it, because it's the kind of thing out of which you can get a lot of publicity. Various students offered to help Pat Palmer and me carry out our project Ozma II, but we felt that the experiment was unlikely to substantially advance their scientific careers.   The main point is that the satellite will do the search. No one will do it from the ground, it's too tedious.

Cyril Ponnamperuma: You have talked about the active searches. Has anyone taken a look at data that has been recorded at various observatories and seen if there is any interesting signal that is sticking out?

Tarter: I am aware of two such searches of radio astronomical data.

Ponnamperuma: Can this be extended for less than $100 per star?

Tarter: Do you have enough graduate students?

Zuckerman: Most radio astronomers tend to throw out records that contain the kind of signals that we are interested in here.

3

An Examination of Claims That Extraterrestrial Vistors to Earth Are Being Observed

Robert Sheaffer

Many members of the general public, and some academic scientists as well, maintain that at least some UFO sightings result from the activities of extraterrestrial visitors. Recent polls show that approximately 57 percent of the public believes that UFOs are "something real" as opposed to "just people's imagination." The figure rises to 70 percent belief for those who are less than thirty years old (Gallup, 1978), and have thus lived their entire lives in the age of television. UFO belief is not found predominantly only among the uneducated. A 1979 poll of its readers by Industrial Research and Development magazine shows that 61 percent believe that UFOs "probably or definitely exist," a figure that rises to over 80 percent for those applied scientists and engineers under age 26. "Outer space" is the most widely-held explanation of their origin.

It is obviously true that even if the reality of UFOs were somehow to be fully established, it would not prove the reality of extraterrestrial visitors. UFOs could possibly be, for example, some poorly-understood atmospheric phenomenon, or the result of some secret terrestrial technology, or even a life form or natural phenomenon which lies totally beyond the scope of present-day science. But in the public mind, the subject of UFOs is inextricably linked with the idea of extraterrestrial intelligent life, and since ETI is the subject matter of this conference, I will henceforth adopt the popular usage of terms, and examine UFO reports in the context of the evidence they purport to contain concerning extraterrestrial visitors.

Although even the strongest proponents of the reality of UFOs concede that the vast majority of reported UFO incidents are the result of the misidentification of conventional objects, they maintain that after the elimination of this noise there remain a reasonable number of observations of UFOs by reliable observers which can only be explained in terms of alien visitors.

However, when we take into account the obvious fallibility of human eyewitness testimony, it is not surprising that there would be a very small residue of supposedly "unexplainable" cases. The police do not achieve a 100 percent solution rate of armed robberies or hit-and-run accidents, yet no reasonable person would suggest that this proves that alien beings have been robbing banks and running down pedestrians. Allan Hendry of the Center for UFO Studies recently published a comparison of the more than 90 percent IFOs — Identifiable Flying Objects — reported to the Center over a one-year period, with the less than 10 percent of the reports which apparently defied identification (Hendry, 1979). His results were truly remarkable. He found that there was no significant statistical difference between the control group of misidentified objects — the IFOs — and the supposedly real UFOs. The two categories were virtually indistinguishable in terms of the duration of the occurrence, the time of day, the age and sex breakdowns of the witnesses, their occupational background, and their previous UFO involvement and interest. (It may be quite significant that both groups, the IFO reporters and the UFO reporters, reveal a degree of prior UFO interest which seems orders of magnitudes higher than that for the general population, although we have no data at this time to make an exact comparison.)

Hendry's data also reveals that of the more than two hundred reports he received that can be unquestionably confirmed as sightings of nocturnal advertising aircraft, more than 90 percent of the witnesses described not what was perceptually available to them, but instead imagined that they saw a rotating, disc-shaped form. 10 percent of these witnesses imagined that they could also see a dome on top the nonexistent "saucer" (Hendry, 1978). Hendry's quantitative evaluation of UFO data has confirmed what many of us have suspected all along: that there is no significant statistical difference between IFO and supposed UFO reports, that raw reports as received by investigative groups are frequently filled with gross errors of observation, and that in a small percentage of cases these inaccuracies are so overwhelming as to totally preclude any rational explanation when the report is taken at face value.

When subjected to a careful and detailed investigation, many of those cases which are frequently cited by UFO proponents as being the most convincing turn out to be readily susceptible to conventional explanation. Perhaps the best-known UFO incident on record concerns the supposed "UFO abduction" of a New Hampshire couple, Barney and Betty Hill. While driving through a sparsely-populated region of the White Mountains on the night of September 19-20, 1961, they allegedly encountered a mysterious aerial object which supposedly interfered with their conscious recollections, and somehow caused them to "lose" two hours from their journey. When the Hills later underwent psychiatric treatment by Dr. Benjamin Simon, a noted Boston psychiatrist, they both told under hypnosis similar stories of being supposedly "abducted" by humanoid alien creatures, and being given a physical examination aboard a mysterious craft. The Hills' account has been the subject of a best-selling book, a made-for-TV movie, was serialized in a major national magazine, and has been cited dozens of times by supposedly scientific UFO investigators as evidence of the reality of alien visitors.

What the UFO proponents nearly always fail to mention, however, is that the Hills' psychiatrist, Dr. Simon, has unambiguously stated on several occasions that, in his professional judgement, the "UFO abduction" story represents a fantasy, and not a real event (Klass, 1974). Mrs. Hill had originally experienced a series of dreams whose content was essentially identical to the story she later told under hypnosis. She described these dreams to anyone who wanted to listen – there were many who did – and her husband, of course, heard her narration many times. Hypnosis is not, we must remember, a road to absolute truth. A person suffering from a delusion might recite that delusion under hypnosis with great sincerity. I should point out that Mrs. Hill's description of the supposed configuration of the UFO and a star near the moon that evening sounds remarkably similar to the moon/Saturn/Jupiter pattern that actually existed. Had a genuine UFO been present, she would have seen three starlike objects in the vicinity of the moon: Jupiter, Saturn, and the UFO. But she saw only two.

Much has been made of an "alien star map" that Mrs. Hill reportedly saw aboard the UFO, and sketched afterwards. Some persons, including a few with impressive academic credentials, have endorsed an analysis purporting to show that the pattern of dots drawn by Mrs. Hill can be uniquely identified as a carefully-selected subset of the nearby stars as seen from a certain perspective outside our solar system, chosen on the basis of parameters that would make them prime candidates for supporting habitable planets (Dickinson, 1974). However, I have elsewhere published an analysis of the inconsistencies and ad hoc procedures required to achieve this supposed match (Sheaffer, 1981). Astronomers Steven Soter and Carl Sagan have noted that, in the absence of a grid of lines drawn in to suggest the relationship, no two patterns could be more dissimilar than the Hill sketch and the main-sequence star pattern (Soter and Sagan, 1975). Perhaps worst of all, there have been three mutually-exclusive purported "identifications" of stars supposedly represented in the Hill sketch published so far, each one having one or more features said to argue in its favor.

In recent years, Mrs. Hill claims to have discovered a UFO "landing spot" in New Hampshire, where she goes as often as three times a week to watch the UFOs land. She claims that the aliens sometimes shoot down beams at her, including one that "blistered the paint" on her car (Hill, 1978). On other occasions, she alleges aliens to have peeped into bedroom windows, and to have gotten out of their parked saucer to do calisthenics before getting back in again (Sentinel, 1978). UFO investigator John Oswald, who is certainly not a debunker, accompanied Mrs. Hill to the "landing spot." He reported upon his return that Mrs. Hill was "unable to distinguish between a landed UFO and a streetlight." He nonetheless still believes that she was abducted by aliens in 1961 (Skeptical Inquirer, 1978).

Another famous UFO incident concerns the supposed near-landing of a UFO on a farm near Delphos, Kansas, in 1971. This case was selected as the most impressive out of a collection of more than a thousand incidents by a panel of UFO researchers holding Ph.D.'s, headed up by astronomer J. Allen Hynek. Not only was there the testimony of apparently credible witnesses, but a whitish powdery ring, whose origin and strange properties seemed to defy analysis, was to be seen on the ground where the UFO allegedly had hovered. Skeptical UFO investigator Philip J. Klass travelled to the site, and interviewed the witnesses. He documented the farmer making several wild-sounding claims which were later shown to be false (Klass, 1974). The supposedly-mysterious ring, which was for several years touted as convincing UFO evidence and to which was attributed totally bizarre physical and physiological properties, was recently disclosed with utterly no fanfare to have been finally identified by a French biologist as a growth of the fungus-like organism Actinomycetaceae, of the genus Nocardia. Information scientist Jacques Vallee suggests, however, that high-energy stimulation from a UFO may have triggered the growth of the Nocardia (Vallee, 1975). The credibility of this case is further eroded by the claims of the principal witness, the farmer's son, that the UFO was responsible for virgin births among animals on the farm, and that he later sighted "the Wolf Girl", who reportedly ran across the moonlit field on all fours "faster than anything human can run" (Sheaffer, 1981).

I could go on to cite several dozen more UFO cases, similar to these, which have risen to prominence. But as this is to be a short paper, the above will have to suffice. I think that all reasonable persons will have to agree that the credibility of the cases described above is essentially nil. The fact that cases like these are touted in pro-UFO writings and endorsed by the few pro-UFO scientists tells us a great deal about the lack of critical thinking which unfortunately is the norm in UFO circles. It also tells us a great deal about the credibility of the numerous lesser-known UFO cases, which must presumably be even less.

For centuries, one of the cornerstones of scientific methodology has been the principle commonly known as Occam's Razor: "Essences are not to be multiplied beyond necessity," or in the vernacular, extraordinary hypotheses are not to be invoked until all ordinary ones have been conclusively eliminated. For example, Occam's Razor would have prevented any truly scientific UFO researcher from writing that a fungus growth in the vicinity of an alleged UFO landing was likely caused by energy emitted from the UFO, unless it could be convincingly demonstrated that such a fungus growth is virtually impossible in the absence of a UFO.

When the evidence offered in favor of the reality of UFOs is critically examined according to the dictates of Occam's Razor, it is clear that there is little or nothing remaining which merits scientific scrutiny, because there is simply no UFO evidence yet presented that is not readily attributable to prosaic causes. The great bulk of what has been offered as evidence for the reality of UFOs consists of unsubstantiated statements by one or more witnesses. We have already noted the observational errors which are rampant in UFO reports. Recent experiments by psychologist Elizabeth Loftus underscore the inherent unreliability of unsubstantiated human observation and recollection (Loftus, 1979). Furthermore, Hendry's data contains the somewhat surprising result that observations reported by multiple witnesses are in fact slightly less reliable than those by a single witness (Hendry, 1979). Perhaps the psychologists can give us some insight as to why that is the case.

Many purported UFO photographs have been published, but in every instance, either the object is insufficiently distinct, or the circumstances surrounding the incident are insufficiently credible, as to render the photograph unconvincing as evidence for anything. Occasionally, tangible effects of a UFO's presence are reported — the so-called "Close Encounters of the Second Kind." But again, nothing has occurred that convincingly rules out prosaic causes. Electromagnetic interference effects have been attributed to the supposed proximity of UFOs. But such effects are sporadic, curiously localized, and always disappear without a trace before they can be verified. In instances where permanent magnetic imprints would be expected if the report was accurate, none have been found (Condon, 1969). The permanent and tangible effects attributed to UFOs consist almost exclusively of things like broken tree branches, fungus-like rings or growths on the ground, and the death of cattle under unusual circumstances.

When the only evidence one can accumulate is of such dubious character, Occam's Razor leaves one no choice but to reject any and all remarkable explanations for the supposed UFO phenomenon, and attribute it to prosaic causes. The "null hypothesis" has not been excluded.

Over the past thirty years many short-lived or rare phenomena have successfully been photographed or detected unambiguously by scientific instruments. Yet UFOs seem to have an utterly infallible ability to avoid unambiguous detection. Alert photographers have captured such brief, unexpected events as commercial airliners falling from the sky, and brilliant daylight meteors. Yet where is the corresponding unambiguous photograph of the often-reported daylight disc UFO? This lack of evidence is all the more puzzling when it is remembered that there is no shortage of UFO reports from major populated areas. We have supposedly authentic close-encounter cases from such places as the suburban areas of Chicago and Pittsburgh, the crowded Santa Ana Freeway near Los Angeles at noontime, and even near the New Jersey entrance of the George Washington Bridge into Manhattan.

Several years ago I coined the term "jealous phenomena" to denote those alleged but never-proven phenomena which have the ability to flawlessly play peek-a-boo with reality. They are defined as those phenomena which _always_ manage to slip away before the evidence becomes too convincing. Joining UFOs are such other jealous phenomena as ESP, Bigfoot sightings, psychic spoon bending, and the Loch Ness monster.

The astonishing elusiveness of the supposed UFO phenomenon, and the undeniably dream-like character of many of the reports, have caused even many advocates of the reality of UFOs to abandon the hypothesis of extraterrestrial visitors, in favor of even more bizarre hypotheses. J. Allen Hynek has suggested that UFOs may originate in some as-yet unknown parallel plane of existence, or may represent non-physical visitations from elsewhere. In a 1976 interview he explained that "perhaps an advanced civilization understands the interaction between mind and matter − in the manner of the Geller phenomena, for instance − much better than we do. Perhaps somebody out there is able to project a thought form and materialize it down here, a la 'Star Trek'. There are other planes of existence − the astral plane, the etheric plane, and so forth" (Hynek, 1976). Jacques Vallee has noted the remarkable similarity between reports of UFO occupants, and reports of fairy sightings of an earlier age. His most recent of many UFO hypotheses is that UFO sightings are caused by some secret human organization for as-yet unknown purposes of deception, using secret high-technology psychic devices (Vallee, 1979). He does not attempt to reconcile this latest interpretation with his earlier hypothesis that sightings of UFO occupants represent the same phenomenon as earlier sightings of fairies, and hence these alleged secret psychotronic devices must have been in existence at least since the Middle Ages (Vallee, 1969). Popular UFO writer John Keel attributes the UFO phenomenon to beings he calls "ultraterrestrials," who he says are "our next-door neighbors, part of another space-time continuum where life, matter, and energy are radically different from ours" (Keel, 1970).

Thus if you, the listener, have always found it next to impossible to give any credence to the extraterrestrial hypothesis for UFOs due to the extreme difficulty of wide-ranging interstellar travel and the implausible character of the reports themselves, it may come as a surprise to learn that there is a significant number of UFO proponents who have recently come to share your opinion. But, unfortunately for this avant-garde wing of saucerdom, their ill-defined ideas about "parallel universes" and "psychic interaction" cannot in any way hope to face up to Occam's Razor.

Another factor which argues against the extraterrestrial or other anomalistic explanation of UFOs is the fact that the percentage of supposed unknowns does not vary with the overall level of sightings. The percentage of supposed unidentifieds is not an invariant, as some UFO researchers have stated, but the percentage varies according to factors other than the overall level of UFO sightings. As an illustration, let us consider some data from the records of the U.S. Air Force's Project Bluebook. 1963 was an extremely quiet year for UFOs; 1966 was a year of great excitement, with nearly three times as many total reports. Yet the percentage of supposed unidentified in both years was approximately three percent. The number of sightings in 1968 was more than double the number of the following year, yet the percentage of supposed unexplaineds was exactly the same, seven-tenths of one percent. Sightings increased by 50 percent from 1956 to 1957, yet the percentage of supposed unknowns stayed virtually the same. Sightings jumped a dramatic 788 percent from 1951 to 1952, and while the percentage of supposedly genuine UFOs increased from 12.6 percent to 19.3 percent, that is a much smaller change than would be expected from the nine-fold increase in sightings. Indeed, one suspects that the Air Force's investigators may have simply been overwhelmed by the volume of reports in 1952 and were unable to investigate them adequately, because if we rely instead on the revised analysis of the Bluebook files recently published (Hynek, 1977), we find that despite the 788 percent increase in reports from 1951 to 1952, the percentage of supposed unidentifieds was essentially the same.

This is a most puzzling factor for the following reasons: suppose that the unidentified UFO reports represent sightings of alien space-craft. Then when the number of genuine UFO sightings triples in a given year, it is presumably because the ones that are here have become three times as active. One would expect the percentage of supposed unidentified to go up dramatically, as the signal to-noise ratio improves, but they do not. Why should a UFO flap cause people to also report Venus and weather balloons as UFOs at three times the previous rate? How did the Venus-spotters determine so accurately the rate of increase of genuine UFO activity that they were able to keep the IFO/UFO ratio essentially unchanged? Or, looking at it from the other side, suppose that the rate of sightings of genuine alien spacecraft does not change significantly from year to year, but that the remarkable "UFO flaps" as we know them are manifestations of a mild form of mass-hysteria. If so, one would expect that the percentage of genuine UFOs would decrease dramatically during a "flap" as the noise increases but the signal remains the same. Yet this does not happen. Do the supposedly genuine alien spacecraft therefore deliberately increase their activities by the appropriate factor to keep the IFO/UFO ratio essentially unchanged?

I suggest that the most straightforward explanation of the above dilemma is that the signal-to-noise ratio in UFO reports is exactly zero, and that the apparently unexplainable residue is due to the essentially random nature of gross misperception and misreporting. This would explain why, no matter how many or how few UFOs are reported, the IFO/UFO ratio does not vary accordingly. If the supposedly genuine UFOs represent nothing more than that minority of cases in which the "random noise" factor in the human perceptual apparatus becomes so large as to totally overwhelm the original stimulus for the observation, then this would explain why the percentage of supposedly real UFOs is essentially independent of rises and falls in the overall number of reports.

Of course, none of the above factors in any way proves that extraterrestrial visitations are not now taking place. Such a proof is impossible. One could no more prove the non-occurrence of such a thing than one could prove that there are no dinosaurs left alive on earth, or that witches do not fly on broomsticks. But what has been established, in my judgement quite convincingly, is that there is no evidence in favor of extraterrestrial visitations that comes anywhere close to establishing that claim as a viable scientific hypothesis. To paraphrase Samuel Johnson, we have no other reason for doubting the existence of extraterrestrial visitors than for doubting the existence of men with three heads: very simply, we do not know that there are any such things. As has been said many times, extraordinary claims require extraordinary proof, and the burden of proof always rests on the person who asserts the existence of an anomaly. Until such proof is supplied, the null hypothesis remains unrefuted, and the existence of extra-terrestrial visitors is not in any way suggested or established.

REFERENCES

Condon, E.U. (editor), (1969), Scientific Study of Unidentified Flying Objects. Bantam, New York. 100; 282 (case 12); 380 (case 39); 749.
Dickinson, T. (1974). Astronomy 2 (number 12), 4.
Gallup, G. (1978). Opinion poll, May, 1978.
Hendry, A. (1978). International UFO Reporter 3 (number 6), 6.
Hendry, A. (1979). The UFO Handbook. Doubleday, New York. Chapters 2-8, 20.
Hill, B. (1978). UFO Report, January.
Hynek, J. (1976). Fate, June.
Hynek, J. (1977). The Hynek UFO Report. Dell, New York. 253-267.
Keel, J. (1970). UFOs Operation Trojan Horse. Putnam, New York, Chapter 15.
Klass, P.J. (1974). UFOs Explained. Random House, New York. 253; 312-332.
Loftus, E.F. (1979). American Scientist 67, 312.
Sentinel (1978). News story of June 27. Centralia, Illinois.
Sheaffer, R. (1981). The UFO Verdict. Prometheus Books, Buffalo, New York. Chapters 5, 18.
Skeptical Inquirer (1978). Volume 3 (number 1), 14.
Soter, S. and Sagan, C. (1975). Astronomy 3 (number 7), 39.
Vallee, J. (1969). Passport to Magonia. Regnery, Chicago.
Vallee, J. (1975), The Invisible College. Dutton, New York. Chapter 1.
Vallee, J. (1979). Messengers of Deception. And/Or Press, Berkeley, California.

DISCUSSION

Sebastian von Hoerner: I am frequently asked, how do I explain that so many people, and so many well educated people, see flying saucers? Although I can't explain that, I can mention a similar example. A few hundred years ago instead of seeing flying saucers many people, and even well educated people, used to see the Devil. Martin Luther threw an ink bottle at the Devil. I can't explain that either. But it shows that people are used to seeing things, and it comes in waves. Now it is flying saucers. I don't know what it will be in a few hundred years.

Unidentified speaker: But it is also true that people used to see meteorites, and meteorites really exist.

Unidentified speaker: Meteorites do not run away.

David Eichler: The most impressive incident that I have heard on the Boston radio was sometime between 1973 and 1976. Forty-five thousand people in a football stadium in Louisiana all jumped up, and many started yelling "UFO".

Sheaffer: Yes, I think that that one was solved in about three days. It was a lighter-than-air balloon.

Unidentified speaker: We didn't read about that in the paper.

Sheaffer: That's right. They only tell us about the unexplained portion. When the explanation comes along they don't bother to print it.

4

The Likelihood of Interstellar Colonization, and the Absence of Its Evidence

Sebastian von Hoerner

According to several estimates, about half a percent of all stars may have a planet similar to our Earth, but on the average about four billion years older than Earth because our Sun is not an old star, and star formation was most productive in the early times. Regarding the origin and evolution of life, our own case is at present the only instance of life we know of. Are we permitted to generalize this single case? Can we do statistics with n=1? As I have mentioned earlier, the laws of <u>statistics</u> say that n=1 yields an estimate for the average, but none for the mean error (which would need at least n=2). This means: assuming us to be average has the highest probability of being right, but we do not have any indication of how wrong this may be. Leaving statistics and arguing by <u>analogy</u>, we may add that most things in nature do not scatter over too large a range, up to a few powers of ten, mostly. Thus, the best we can do is to assume that we are average, but to allow for a wide (but not infinite) error of this assumption. If we now generalize our own case, then life in our Galaxy would have started on about one billion planets several billion years ago. And arguing by <u>extrapolation</u>, we should expect this life to have developed meanwhile extremely far beyond our own present state. However, nothing but <u>wild guesses</u> can be made regarding the direction, the range, and possibly the termination of such far-out developments. What activities should we then expect that would be visible to us? And mainly: Why don't we see any? This is a great, tantalizing puzzle.

Looking at our own present activities brings us back to Frank Drake's old question: "Is There Intelligent Life on Earth?" Our own large-scale activities are mostly self-destructive, with a world-wide arms race of 400 billion dollars per year, which is 100 times the budget of NASA or 400 times that of the National Science Foundation. All the big nations spend about 25-50 percent of their total governmental budget, or about 5-10 percent of their Gross National Product, on what each side calls defense. It is clear that this enormous arms escalation cannot go on forever. We shall either blow up, sooner or later, or we must redirect this tremendous work force to peaceful endeavors, hopefully to the exploration and exploitation of all the many resources of our solar system and to large dedicated SETI projects.

Provided we do have a future, let us guess some of our future activities. We are now using up our terrestrial resources at an alarmingly fast and still increasing rate, and within a few more generations we must develop, introduce and enforce rather elaborate and expensive recycling systems for most of the metals and some other minerals; but no recycling system can ever work a hundred percent. We have just now begun to feel an energy crisis, the really dramatic part of which is still ahead of us. But if demands are increasing and supplies are running out, why on Earth should we stay?

The following lectures will show that large self-sustaining, growing and multiplying colonies throughout our solar system can be made, for mining and production, even with our present beginner's technology and with less money than the suicidal arms race. After some large original investments, the colonies will become quite vital and profitable when our resources on Earth run out. These colonies will grow with their own babies and grandchildren, and, after more and more generations out in space, their people will feel less and less of their originally strong ties to the home planet Earth, probably up to some Declarations of Independence.

Larger groups of thousands of volunteers may decide to take off in huge "mobile homes" on interstellar trips lasting many generations, finally colonizing the planets of other stars. And after a certain settling time on such a planet, the same cycle may repeat, leading again to mobile homes on another interstellar trip to the next stars and their planets. In this way we would have started a wave of stepwise colonizations, finally covering the whole Galaxy from one end to the other, and every nice planet in it. The complete Galactic colonization could well be finished within some ten million years or less, even with our present limited knowledge of physics and technology.

And here we have our great puzzle. Such a wave of colonization could have been started by any one out of the billion early civilizations in our Galaxy. Our Earth should have been colonized long ago, and we ourselves should be the descendants of some early settlers, and not the homegrown humans that we certainly are.

Should we really assume such an urge for interstellar colonization, is this reasonable and justified? First, it is only necessary that at least one civilization felt the urge strong enough to get it done, one out of a billion potential ones. Furthermore, quite in general, life shows a strong tendency to fill out every possible niche, from dry deserts to cold polar regions, from swamps to bare rocks, from caves to the tops of mountains. Life has started in the water, it has conquered first the land and soon the air; it begins right now to conquer nearby empty space, and so it may quite naturally proceed to conquer interstellar distances as well.

What enables or triggers the larger steps of this development? The decisive milestones of evolution are set by introducing and exploiting new ways of information handling. All life, self-reproductive life, began with the genetic code, which is a most ingenious way of storing and reproducing all the information needed for storing and reproducing that very information plus whatever is needed for growing the whole organism surrounding it, including information about maintenance and repair and behavior. The next large step, the development of higher life, was made possible by growing a nervous system with a brain as its main office of command, where incoming informations are evaluated, memories are stored, and outgoing instructions are given. The third large step, our whole human culture, is based on the development of speech about a million years ago. It has been drastically enhanced by the introduction of script, and now it begins another revolution by using cybernetic means. And so this whole evolution should quite naturally proceed to a large-scale network of interstellar communication, involving all the many members of the Galactic Club. And again we have our puzzle: that all this should actually have been done long ago, that the whole Galaxy should be teeming with life, that "empty" space should be bristling with messages and probes, some of it obvious in many ways, whereas we have not yet found any evidence of any extraterrestrials.

Many reasons against colonization have been mentioned for explaining the puzzle. For example: self-destruction of technologies, biological degeneration, stagnation by over-stabilization against crises, complete change of cultural interests, space technology never becoming cost-effective, a repetition time for traveling and settling of more than a million years, colonization turning from an organized procedure into a random walk. All of these reasons could very well hold in some cases, maybe even in most cases, but hardly with no exception at all in a billion. What matters is the far-out tail of a highly populated distribution.

A possible conclusion then is that we are actually alone, by some reason not yet understood. Or could it just be that one in a billion of nice habitable planets has been overlooked or neglected by the colonizers? And regarding all the other evidence to be expected: maybe we do not look for the right thing, or we do not understand what we see? The great puzzle is still unsolved.

REFERENCES

Ball, J.A. (1973). Icarus 19, 347.

Bova, B. (1963). Amaz. Fact. Sci. Fict. 37, 113.

Bracewell, R.N. (1974). The Galactic Club. W.H. Freeman, San Francisco.

Cameron, A.G.W. (editor), (1963), Interstellar Communication. Benjamin, New York.

Dole, S.H. (1970). Habitable Planets for Man. Elsevier, New York.

Dyson, F.J. (1968). Physics Today 21, 41.
Hart, M.H. (1975). Quarterly J. Royal Astronomical Soc. 16, 128.
Hoerner, S. von (1975). J. Brit. Interplanetary Soc. 28, 691.
Hoerner, S. von (1978). Naturwissenschaften 65, 553.
Jones, E.M. (1976). Icarus 28, 421.
O'Neill, G.K. (1974). Physics Today 27, (number 9), 32.
O'Neill, G.K. (1977). The High Frontier. William Morrow, New York.
Papagiannis, M.D. (1977). International Conference on the Origin of
    Life, Japan. Astron. Contrib. Boston Univ., Series II, (number 61).
Sagan, C. (editor), (1973). Communication with Extraterrestrial Intel-
    ligence. MIT Press, Cambridge, Mass.
Sampson, A. (1977). The Arms Bazaar. Viking Press, New York.

DISCUSSION

Eric Jones: A brief comment on the zoo hypothesis. One factor that needs to be considered is that stars do not march around the galactic center in a nice orderly fashion. They have some random motions. So that if you have a region in which an ancient civilization sets up a wilderness preserve, it will soon find the region dispersed over a substantial fraction of the galaxy.

David Eichler: Cosmic hazards such as supernovae might make most of the galaxy uninhabitable.

Jill Tarter: If colonization is such a natural consequence, and N is so large, then why doesn't competition have to be included in the expansion calculations?

Von Hoerner: Well, in that case there will probably be some sort of galactic club and things will be organized in a nice fashion. Those people who are as destructive as we are never make it — they destroy each other before they can go anywhere. We have to solve the question of peace first, and then maybe we have the possibility of going somewhere else.

Gerald Feinberg: The title of the symposium, "Where are they?" can be interpreted in at least two ways. One is, "why aren't the extraterrestrials here", and the other is, "why don't we see evidence of them out in the sky". Now on the second question, something you said suggested that you thought that there was evidence against there being advanced extraterrestrials out there. Could you elaborate on what you consider that evidence to be?

Von Hoerner: We have looked around and made these radio searches. That is only a very small beginning in terms of effort and money. But on the other hand, all astronomers are now more or less aware of the problem, and when they do normal astronomy they sort of have an open eye for things which look unnatural.

Feinberg: But you have to have an idea of what to look for.

Von Hoerner: No, you don't. If a native from some tribe comes for the first time to a city he will realize that this is artificial, although before he didn't know what to expect.

Feinberg: But I don't think that is a good analogy. One is talking about a situation where you are looking at something far away and you don't really know what should be there in the first place. There are things that we have discovered, such as quasars, that we don't understand by the natural laws of physics.

Von Hoerner: A counterexample is pulsars. When they were discovered one of the first ideas was little green men.

Feinberg: But that's still only one item. Morrison and Cocconi had one good idea about looking for radio signals, and that's it. That's what we have been living off of for the last 15 or 20 years. I suggest that maybe we need some more thought on what intelligent extraterrestrials might be doing — not designed to communicate with us, but just to further their own purposes.

Von Hoerner: Yes, we have eavesdropping, for example.

Michael Hart: A Kardashev type II civilization will give off infrared radiation in large quantities.

Feinberg: Well, maybe two good ideas, although I think that the details of what you should look for to detect those have not been completely clarified.

Von Hoerner: With respect to having a good or the best search method, I would like to remark that first you need a method that is possible. Now we do have a possible method: that is radio. It may not be the best method. But whether we have the best one is something we never will know — not now, not in a thousand years, not in a million. Since that is always the case, we might as well start now.

James Oberg: A science fiction theme is that only the belligerent people will get out and explore. The calm and peace-loving people will sit around in circles, chant the theody, and they will never go anyplace. It's the scrappy ones who barely get through wars by the skin of their teeth who will get out into space, and they will carry that with them. What we will observe is their star wars. We will watch and see the X-ray bursts and we will see those things going on which are in fact not their industrial activities. Man's greatest efforts have been military. The same may be true elsewhere in the galaxy.

5

Preemption of the Galaxy by the First Advanced Civilization

Ronald Bracewell

In one school of thought it is customary to begin discussions of galactic life by appeal to Drake's equation and then to proceed to a detailed examination of the numerical magnitude of one or more of the string of factors whose values have to be estimated. An example of this procedure is furnished by Michael H. Hart's analysis in which he concentrates on the probability that 600 or more nucleotides might line up in the right order; then he proposes that one of the factors may be very much less than 10^{-30}. Of course, 10^{-30} is already very small and if included as a factor in almost any expression having to do with the physical universe will cut the product down to negligible size. In this application the conclusion is that the number of technological civilizations independently arising in a galaxy is very much less than one. Well, this may be an excess of zeal, and many of those addicted to the use of Drake's equation would, in similar circumstances, have arranged for the product to emerge with an order of magnitude around unity because, after all, a calculation condemns itself if it seriously contradicts the possibility of the one technological civilization we know about, namely our own.

But is Drake's equation correct? It seems that it suffers from oversimplification – surely at least one plus-sign ought to be there. This matter was taken up at greater length in connection with a meeting of the International Astronautical Federation(1) and the conclusion is that the equation has to be generalized. At the very least, more than one route to life and certainly to intelligent life, has to be allowed for. As the cited paper states, "One would not attempt to estimate N_f, the number of fish in the sea, by multiplying an estimate of R_f, the rate of formation of fish, by L_f, the average longevity of a fish. There are too many kinds of fish to make this simplification useful." There are other troubles with the scientific foundations of Drake's equation.

My inclination is to admit that the plurality of technological civilizations is an open question and to follow up the different logical possibilities with a view to discerning courses of action that we might take. Thus, it is conceivable that the Galaxy teems with life – in that case radio listening makes sense and will soon provide confirmation. Happily that kind of action is already under way. But life is conceivably much more rare. If life is too sparse, radio listening might be fruitless, but there might be something else we could do. As is well known, a proposal was made in 1960(2) that in such a case, referred to as Case II, we should be alert to the possible appearance of a messenger probe. At that time it seemed sensible, on a basis of cost effectiveness, that such an alien messenger probe ought not to bear the penalty of having to make a planetary landing on unfamiliar terrain but should simply go into orbit in the habitable zone of the sun and explore the radio spectrum for indications of technological activity. But now it seems to me that, under Case II, some principle of equipartition of effort ought to apply. A move toward such a principle already underlies the Cyclops philosophy where, however, it is understood that the alien civilization maintaining the powerful radio beacon over millennia is making a contribution that far outweighs the terrestrial contribution, which is merely to construct and man a listening post for perhaps a decade; even so the Cyclops design would strain national resources.

The equipartition principle, then, means that the participants make efforts that, in terms of their capacities, are comparable. Applying this principle to contact by messenger probe, we ask ourselves, what efforts is it that Earth could make that exceeds the role of mere alertness, as originally suggested, and that substantially enhances the effectiveness of the probe. One answer is that we could relieve the probe of the need to orbit the sun by maintaining a solar system space watch for a probe that may simply be flying by. Such a probe would approach at a great speed and we would have to detect it in the depths of space and act quickly. A velocity of several tens of kilometers per second could be expected – probably something in excess of the escape velocity from the solar system. Even so, as we know from the behavior of comets, such as Halley's comet, that have come from great distances, the time of passage through the inner solar system is many days. This should be sufficient to extract the message content from a cooperative probe. Whether there would be a bonus to be gained from the greater effort on our part that would be entailed in a rendezvous with the probe, and whether such a rendezvous would be depended upon by the designers of the probe, seems doubtful to me because the lodging of a message is paramount and can be done by radio without need of physical contact, fascinating though that could be.

Under this more equitable sharing of effort it would not be necessary to equip the probe with the very substantial retrorocket needed for halting here and as a result more probes and more reliable ones could be launched under whatever constraints were operative. The above discussion is to illustrate my theme of identifying separate courses of action appropriate to various logical possibilities as to density of technological life. Of course, in both cases so far mentioned technological life exists: in Case I it is dense and in Case II it is less dense.

There is another good illustration of this action-oriented theme which applies even if life is sparse, in which case one may at least seek planets. There are various ways of searching(3) all of which require to be developed simultaneously. A scheme using an apodized space telescope worked out by B.M. Oliver is very promising and needs to be followed up. Another scheme proposes an interferometer(4,5) spinning in space and makes use of infrared radiation from the hypothetical planet because such radiation is 10^5 times stronger relative to that from the parent star than is the case with visible light. Both the space-based schemes involve advanced technology that does not yet exist, but there are no apparent fundamental limitations, and the prospects of technical success are favorable. In this illustration the action indicated is development of appropriate space technology. We are reminded that, where technology (as distinct from science) is involved, certain decisions are better postponed pending the outcome of preliminary studies and necessary experimental development.

Among the logical possibilities is one that has not been much favored but which in my opinion requires equal attention, again with a view to discerning appropriate action. I take the orthodox scientific position of not making up my mind until there is evidence. To some people such fence-sitting is uncomfortable — they like to make up their minds and to be known as decisive and nonstodgy people. Way back in 1964 I wrote a well anthologized chapter entitled "Are We Alone?" and it wasn't long before Walter Sullivan took a liking to this phrase and brought out a book entitled, "We Are Not Alone." Well, I don't know how he knows that we are not alone. We very well may be alone, but this has not been a favored hypothesis even though it leads to a most interesting discussion. It is Case IV. (It should be mentioned that Case III refers to a situation where technological life in the Galaxy is so sparse that under secular equilibrium conditions the longevity of a civilization is less than the round trip travel time of electromagnetic radiation to the nearest neighbor. The conclusion that longevity has an inverse relation to the spacing of neighbors(2) under equilibrium conditions is now generally accepted.)

Here is the reasoning connected with Case IV. Intelligent life on earth evolved through competition under definite conditions of time and place that might, as far as we know, have arisen at other times or in other phases on earth, but did not. Our arboreal ancestors in Africa developed the precise binocular color vision that we benefit from today, together with the manual dexterity and coordination of hand and eye that are desirable in leaping from branch to branch. On descending to the forest floor our ancestors developed other human skills, tool making and speech, possibly under the demand for weaponry and advance planning for group action fostered by big game hunting. Be these details as they may.

Why didn't a tree kangaroo descend to the Australian savannah, adopt an upright posture, move to a mixed diet that included occasional birds' eggs and lizards and go on to group hunting of giant macropods? Honing his intelligence on internecine warfare with related kangaroo species an intelligent marsupial might have emerged accompanied by no relatives closer than the peaceful grazing kangaroos of today which perhaps might have become the basis of a meat and leather industry. Similarly, why didn't the three-toed sloth descend onto the South American pampas or the raccoon emerge onto the prairies to hunt the buffalo. There are any number of creatures with abilities comparable to those of our arboreal ancestors, many living in geographic conditions not obviously different from what Africa offered. Why haven't they progressed further on the upward path to civilization?

Regardless of the detailed reasons that may have impeded such progress in the past we can now relate what in fact did happen once man passed the threshold of mobility, because he migrated over the whole earth and by his presence now preempts the possibility of future evolution in directions that would compete with his supremacy. Signs of intelligence would bring immediate retribution, and presumably the absence of any gradations between ourselves and the chimpanzee is due to harsh suppression that occurred at some time in the past when the struggle to determine which intelligent strains would survive was still unsettled.

It is not customary in biology to speak as if Nature had a plan but it leads to convenient phraseology. If there was a plan to populate the earth with intelligent creatures it did not depend on trying out different schemes at different times and in different places. On the contrary, Nature's plan was to do a good job in just one place. It took a lot of time. But when a critical threshold was passed that successful progenitor moved out to cover the whole world. Very possibly some further physical evolution took place along the way, but all in the same line of descent — no bears or raccoons were called in once the primate line paid off. So, when the age of world exploration began, men found all the continents except Antarctica and many remote islands to be already populated by intelligent beings — their own relatives.

The fact that the earth is populated with intelligent creatures is not because the many habitable areas of earth were able to foster the evolution of intelligence, it is because one area was the scene of the events; and the rest of the earth is now a habitat of intelligence, courtesy of Africa. Even Antarctica is now inhabited by man.

In time to come, perhaps not too far off, there will be colonies on the moon and in space, comparable with the many colonies now maintained in Antarctica. We shall then have to admit that cislunar space is populated, and then in due course Mars and perhaps other planets and satellites will be. Until recently it was considered quite possible that life might have originated independently on Mars, and probably this possibility is not entirely abandoned even today, but it is now clear that intelligent life will not evolve on Mars. Even if the weather were to improve, the possibility of Mars being left intact for the necessary eons has been preempted by the appearance of man. Mars may indeed come to be populated, but not by Martians.

When this comes to pass we could confirm that Nature's plan to populate the solar system had indeed been to develop a good mobile model in one place and have it migrate to other habitats.

As we can already foresee the first unmanned exploration to nearby star systems it is natural to turn now to speculation about the Galaxy as a whole. Perhaps you think Nature's plan is to have man fill the Galaxy following the pattern of the continents and the foreseeable pattern for the planets. But I do not believe you·can so conclude before asking why it was that the plan worked on earth. There is a very simple mathematical consideration that sheds light on this question, an idea that was first published in the San Francisco Examiner, June 29, 1975 where it could have had no effect on scientists except those on the Berkeley-Stanford axis. The idea is simply that humans could walk from Africa to California, long and hard though the trek may have been, in less time that it would take for intelligence to evolve independently in California or elsewhere and therefore that the key question is, How long would it take .to migrate through the Galaxy as compared with evolution time? The conclusion was: "At moderate speeds of space travel the human race can reach the center of the Galaxy in much less time than the 3.5 billion years that it took the earth to produce man."(6)

This is a fascinating conclusion. It does not say that we will do it but it says that we can do it. It changes our self-estimate from what Cyclops implies, namely that we are inferior, to the view that we may have a noble destiny. Let's follow up this view.

We are profoundly cosmic creatures. The majority of the atoms in our bodies are hydrogen atoms that were created in the Big Bang 15 billion years ago. Most of the mass of our bodies is in oxygen atoms that were created in the generation of stars, now gone, that preceded the formation of our sun. Human protoplasm is continually being created from the lifeless mineral constituents of our planet. We are a subset of the material of the physical universe and now we are confronted by the spectacle that a negligible subset of the universe is, through astronomy, acquiring an awareness, albeit imperfect, of the total universe. How can the part be aware of the whole? But not only that, we are beginning to understand the universe and to bring parts of it under control. At the same time we are converting the inanimate physical matter into protoplasm.

Now we are on the threshold of distributing our protoplasm, and the intelligence and awareness that go with it, back into the cosmic space which is where we came from, from the stars. We are in a real sense part and parcel of the Galaxy, not mere observers, onlookers. Perhaps our role will be to influence the physical evolution. of the Galaxy, bringing to galactic evolution the phenomenon of consciousness which already influences man's own progress and, though as yet geographically very confined, is nevertheless a part of physics. If we are alone, we may be on the threshold of a magnificent destiny.

REFERENCES

(1) Bracewell, R.N. (1979). An extended Drake's equation, the longevity-separation relation, equilibrium, inhomogeneities and chain formation. Acta Astronautica, 6, 67-69.

(2) Bracewell, R.N. (1960). Communications from superior galactic communities. Nature, 186, 670.

(3) Bracewell, R.N. (1979). Life in outer space. Proc. Roy. Soc. New South Wales 112, 139.

(4) Bracewell, R.N. (1978). Detecting nonsolar planets by spinning infrared interferometer. Nature, 274, 780.

(5) Bracewell, R.N., and MacPhie, R.H. (1979). Searching for nonsolar planets. Icarus, 38, 136.

(6) Bracewell, R.N. (1975). Other Voices. San Francisco Sunday Examiner and Chronicle, B3, (June 29 issue).

DISCUSSION

Harlan Smith: The real radiation of evolution is probably still to come, assuming that we make it through the steps that you are describing, because, as you know, characteristics of populations tend to change most rapidly when the breeding population is small. What is stopping evolution from progressing rapidly now is the huge diffuseness of people, so that any mutation tends to get swallowed up rather than have a chance to become established. So the point is that once we do begin to get small populations moving out to the stars, that's where the explosively new radiation of evolution is likely to occur. And even if your vision should prove to pan out, I suspect that it will be something radically different than us characters in this room that will ultimately inherit the galaxy.

Bracewell: I don't believe that, because I think that biological evolution has come to an end. The simple reason that I have long legs attached to the upper part of my body is to enable me to run after large game. But that type of evolution as applied to all parts of the body ceased long ago when I could get an advantage over you not by having longer legs but by having a spear or something that my daddy told me about that gives me an advantage in tactics. So it ceased to be advantageous to have biological evolution once we have vision and technology to pass on. So I don't see any reason at all why people in space ships would evolve, unless it's to have their legs dwindle away entirely.

Gerald Feinberg: How about getting smarter?

Bracewell: We don't need to, you see in fact the brain. . .

Feinberg: Speak for yourself.

Bracewell: The brain has . . .

Feinberg: It's hard to see why human beings as we are in this room are the smartest possible creatures.

Bracewell: You are quite right. We may already have wiped out people who are smarter and nicer than we are. But just as a spear can be viewed as a prosthesis that gives you greater killing power at a distance, so also the computer is something that can extend your brain power; and you don't need to have a brain within your skull that is any better. It won't be very long before we will get plug-in memories. In fact, a piece of paper is an adjunct of a technical kind to what you have to store in your head. I make notes here instead of memorizing. So you know the brain has really stopped moving.

Robert Shapiro: When it comes to evolution, people assume that the old Darwinian mechanisms, which are about the equivalent of our legs, will continue unabated when modern advances in nucleic acid chemistry make it quite clear that, long before we can think of going to the next part of the Galaxy, we will have completely gained the ability to control our shape and entire biological form. Where the race goes after that will be under our conscious control, and not subject to random mutations.

Bracewell: But if you are right, we will dispense with the legs and probably one of the arms, and we will have eyes for looking at TV displays, and one hand left for interacting with the knobs.

Unidentified speaker: Apparently the various fossil hominoids co-existed for a long time. What evidence is there that one so-called more advanced form exterminated the more primitive forms?

Bracewell: There is that fossil which has a broken skull, remember? The evidence is that we have only one kind now. Whereas not very long ago, certainly since the ice age, which is only 10,000 or 12,000 years ago, we had Neanderthal and Cro-Magnon man which were noticely different, sufficiently different that their fossils are identifiable by name. They are not with us now. So it appears that they have disappeared. We could go back a million years ago to all those other sorts of hominoids that you mentioned. They have all completely disappeared.

6
Interstellar Propulsion Systems
Freeman Dyson

There is no lack of propulsion systems available to any creatures which possess some technical competence and a desire to travel around in the galaxy. The following is an incomplete list of propulsion systems which have been suggested and studied by members of our own species.

Group A: Systems which are certainly feasible but are limited to mission-velocities of the order of $10^{-2}c$.

1. Nuclear-Electric. Uses a fission reactor as energy source, and ion-beam or magnetohydrodynamic plasma-jet for propulsion. One can imagine a "minimal starship" using nuclear-electric propulsion, with 10^{-6} g acceleration, a mass of five kilograms per kilowatt (electric) for reactor and radiator, and a mission duration of 10^4 years for voyages of the order of 10 parsecs.

2. Old-fashioned Orion nuclear pulse propulsion, using full-sized fission or fusion bombs. This is also limited to velocities of the order $10^{-2}c$ but can have acceleration of the order of 1 g, giving it much better performance in local maneuvers. (See Dyson, 1968; Martin and Bond, 1979).

Group B: Systems which are probably feasible but require very demanding new technology. These systems should be capable of mission velocities of the order of 0.5c, and mission durations of a few decades for distances of a few parsecs.

3. Laser-driven sails (See Norem, 1969).

4. Pellet-stream propulsion (See Singer, 1980).

5. Direct electromagnetic launch (See Clarke, 1950).

The laser-driven sail system requires a laser with output power of 10^{13} watts, emitted coherently over an aperture of the order of 30 km diameter.

The pellet-stream system uses a transmission-line magnetic accelerator or rail-gun to accelerate solid pellets to moderately relativistic velocities. The ship has the job of intercepting and absorbing momentum from the pellets. If the pellets can be accelerated at 10^4 g, the length of the gun will be of the order of 10^8 km.

Direct electromagnetic launch uses a similar rail-gun to accelerate a space vehicle in one piece. This would be a preferred alternative if the passengers in the vehicle could survive high accelerations.

Group C: Systems which I shall not discuss because I consider them to be inferior either in feasibility or in performance to Groups A and B.

6. Project Daedalus (JBIS, 1978).

7. Bussard Ram-Jet (Bussard, 1960).

8. Celestial Billiards (Ford, 1959).

9. Matter-Antimatter Annihilation Rocket (Steigman, 1974).

For further references to all these systems, see the bibliographies by E.F. Mallove and R.L. Forward (1972, 1977).

The most efficient and economical systems are those of Group B, which have the energy source located in a massive fixed installation decoupled from the vehicle. The vehicle, not needing to carry its own energy source, can be made small and light. These are the only systems which offer travel at relativistic speeds with reasonable efficiency in mass and energy.

Unfortunately, the Group B systems are better at accelerating than at decelerating. Norem has discussed a method of deceleration for the laser-driven sail, using the deflection of electrostatic charge by the interstellar magnetic field to turn the vehicle velocity around through 180 degrees, so that the vehicle can decelerate using the same laser beam that was used to accelerate it. A similarly complicated maneuver has been suggested by Singer for deceleration in the pellet-stream system. These deceleration maneuvers greatly increase the size and complexity of the launch systems and are useless for voyages extending more than a few parsecs from the launcher. So the main conclusion which follows from this examination of interstellar propulsion systems is the following:

WE HAVE THE ACCELERATOR!
HOW ABOUT THE BRAKES?

The Group B systems would provide versatile and efficient rapid transit around the galaxy, provided that there existed a correspondingly efficient and versatile braking system. By a "braking system" I mean a self-contained device by means of which a fast-moving vehicle could slow down and stop, transferring its momentum to the interstellar plasma. I know of only one candidate for a braking system, namely:

Group D: Braking systems.

10. Alfven Propulsion Engine (See Drell et al, 1965). The Alfven Propulsion Engine was discovered accidentally when it turned out that the ECHO I satellite (a large balloon orbiting the earth at 1600 km altitude) experienced a drag force exceeding the expected aerodynamic drag by a factor of about fifty. Drell, Foley, and Ruderman explained the anomalous drag as the effect of electromagnetic coupling between the vehicle and the plasma in the Earth's magnetosphere. The momentum of the vehicle was efficiently transferred to the plasma by Alfven waves traveling along the magnetic field-lines.

The theory of Drell, Foley and Ruderman gives a simple formula for the Alfven drag,

$$D_a = \rho V V_a A, \tag{1}$$

where ρ is the mass density of the plasma, V the vehicle velocity, A the frontal area of the vehicle, and V_a is the Alfven velocity

$$V_a = \sqrt{B^2/4\pi\rho} , \tag{2}$$

with B the magnetic field component transverse to the flight path. The theory was derived from Maxwell's equations and verified by the ECHO satellite only for the case of sub-Alfven vehicle velocity

$$V << V_a . \tag{3}$$

In the earth's magnetosphere, V_a is of the order of 100 km/sec and the theory is certainly applicable. Unfortunately, the Alfven velocity in interstellar plasma is only of the order of 10 km/sec, and the interstellar vehicles will be in the super-Alfven regime

$$V >> V_a . \tag{4}$$

It is an urgent theoretical problem to investigate whether the drag formula (1) still applies in the super-Alfven regime.

Another way of writing the Alfven drag formula (1) is

$$D_a = \sqrt{D_h D_m} \, , \qquad (5)$$

where D_h is the ordinary hydrodynamic drag

$$D_h = \rho \, V^2 A \, , \qquad (6)$$

and D_m is the magnetic drag

$$D_m = (B^2/4\pi) \, A. \qquad (7)$$

In the earth's magnetosphere the true drag D_a is larger than the hydrodynamic drag D_h. I am conjecturing that in the super-Alfven regime the true drag is still D_a, although it is then much less than D_h.

If (1) holds, then the deceleration of a vehicle of mass m is given by

$$m\dot{V} = -\rho V V_a A \, . \qquad (8)$$

The vehicle comes to rest exponentially fast, with a characteristic stopping-time

$$\tau = \frac{m}{\rho V_a A} \, . \qquad (9)$$

A typical interstellar medium has ρV_a of the order 10^{-18} gm cm^{-2} sec^{-1}. To give a stopping-time of the order of a few years, we need a vehicle mass per unit area

$$\frac{m}{A} \leq 10^{-10} \text{ gm/cm}^2 \, . \qquad (10)$$

A mass per unit area of 10^{-10} gm/cm^2 would be absurd for a continuous surface, but it is perhaps not absurd for a braking system. The braking system need not cover its area continuously, but could consist of thin long wires spread far apart. For example, it could carry current in a network of 1-micron-diameter wires spaced 10 meters apart. Detailed analysis is needed to find out whether such a network could be made electrically conducting and mechanically strong enough to decelerate itself by coupling to the interstellar plasma.

If it turns out that interstellar braking systems are feasible, then we have a new way to look for evidence of extraterrestrial intelligence. Look for skid-marks on the road! A vehicle braking from high velocity will leave behind it a long straight trail of hot plasma which should be a source of persistent broad-band radio emission. Radio astronomers interested in CETI should be on the look-out for straight tracks of glowing plasma in the sky. They should also be careful to make sure their signal-to-noise ratio is high enough, so that their tracks do not prove as illusory as Lowell's Martian Canals.

REFERENCES

Bussard, R.W. (1960). Astronautica Acta 6, 179-194.

Clarke, A.C. (1950), J. Brit. Interplanetary Soc. 9, 261-267.

Drell, S.D., Foley, H.M., and Ruderman, M.A. (1965). Physical Review Letters 14, 171-175.

Dyson, F. (1968). Physics Today, (October issue), 41-45.

Ford, K.W. (1959). Los Alamos T-Division Report.

JBIS (1978). J. Brit. Interplanetary Soc., Interstellar Studies Supplement.

Mallove, E.F. and Forward, R.L. (1972). Bibliography of Interstellar Travel and Communication-1972. Research Report 460, Hughes Research Labs, Malibu, California 90265.

Mallove, E.F., Forward, R.L. and Paprotny, Z. (1977). Bibliography of Interstellar Travel and Communication, April 1977 update. Res. Rep 512, Hughes Research Labs.

Martin, A.R. and Bond, A. (1979). J. Brit. Interplanetary Soc. 32, 283-310.

Norem, P.C. (June 1969). American Astronautical Soc., Paper, 69-388.

Singer, C. (1980). J. Brit. Interplanetary Soc. 33, 107.

Steigman, G. (1974). Yale University Preprint.

DISCUSSION

Gerald Feinberg: You get a certain amount of drag just by friction against the protons in the interstellar medium. I wonder how that compares in magnitude with the Alfven drag.

Dyson: Well the ratio is the same as the ratio between the vehicle velocity and the Alfven velocity.

Feinberg: Which one is bigger?

Dyson: The vehicle velocity is larger than the Alfven velocity by a factor of one thousand or so. This means that the mechanical drag is much larger than the Alfven drag.

Feinberg: But then wouldn't that be a fast way of slowing down – by extending out big panels?

Dyson: The mass per unit area has to be very small – a very thin foil. But then a thin foil doesn't stop a proton. So you have to use magnetic coupling. No way to avoid magnetic coupling.

7

Settlements in Space, and Interstellar Travel

Cliff Singer

ABSTRACT

An upper limit to the amount of effort required for interstellar travel by extraterrestrial intelligence (ETI) can be deduced by examining the options available to mankind with forseeable technology. On-board nuclear propulsion should allow travel at up to 1 percent of the speed of light, and recent concepts in remote propulsion may allow much faster travel. The requirements for constructing a self-replicating manned ecosystem from debris in the solar system have recently been examined in great detail, and the propulsion system needed to use such a habitat as an interstellar vehicle is readily determined. However, biologically advanced ETI may have additional options available for interstellar travel, and some of these options have not yet been given sufficiently careful consideration.

INTERSTELLAR TRAVEL AND
EXTRATERRESTRIAL INTELLIGENCE

The success of several proposals to search for extraterrestrial intelligence (ETI) in the galaxy (Cocconi and Morrison 1959, Oliver and

*I would like to thank H. Taylor for his flexibility in administering a National Science Foundation Postdoctoral Fellowship which supported the germination of these ideas several years ago, F. Dyson for pointing out the existence of the work which led to the idea presented in Appendix B, and B. Pieke for reading the manuscript. I also thank E. Carey for typing the manuscript when she could have been drinking tea, and the Princeton Plasma Physics Laboratory for the artwork; other than these services, this work was not supported by the U.S. Department of Energy Contract No. DE-AC02-76-CH03073.

Billingham 1971, Michaud 1979) requires the existence of a large number of technologically competent cultures over a long period of time. For example, to expect to find one ETI within one thousand lightyears in a perfectly efficient search would require about a million ETI in the galaxy, each signaling for a million years. (Or it would require 10^8 ETI signaling 10^4 years, or 10^4 ETI signaling 10^8 years, etc.) Many people have asked why some of these ETI should not have taken advantage of their prolonged technological capability to find a method for interstellar travel and settlement of nearby stellar systems (e.g., Hart 1975, Jones 1976, Winterberg 1979). If the initial problem of interstellar travel and settlement were solved, then it should become progressively easier for daughter settlements to eventually continue the process until every available stellar system in the galaxy (including possibly our own) were inhabited.

The chances of this happening have been discussed extensively in this conference and elsewhere, often with minimal thought given to the physical requirements for interstellar settlement. In particular, it has been argued that interstellar settlement is either impossible (e.g., Purcell 1960, Marx 1973) or absurdly expensive (e.g., requiring trillions of man-years of effort to amass the nuclear fuel needed). These arguments require some attention, especially in light of the possibility that some of the best known advanced propulsion methods (e.g., pulsed pure fusion, Bond et. al. 1978, Winterberg 1979) may be intrinsically unworkable. Given such uncertainties (and more profound ones discussed in Appendix A), it would be naive to expect we can determine the minimum physical requirements for interstellar settlement. However, this does not terminate sensible discussion of this topic. For we can set approximate maximum physical requirements by examining the limited set of possibilities potentially available with application of our present understanding of physics, biology, and engineering. The discussion will therefore now be limited to the possibilities available to mankind in this context.

SPACE HABITATS IN THE SOLAR SYSTEM

Some insight into the possibilities for settlement and propulsion in space is now available as a result of recent work at Princeton, NASA-Ames, and elsewhere (Johnson and Holbrow 1977, O'Neill and O'Leary 1977, Arnold and Duke 1978, Grey 1977) on the prospects for space settlement and space manufacturing in earth orbit in the near future. This work forms a relatively solid base for discussion of at least some of what we may do in the solar system in the next few hundred years. Since discussing our potential for interstellar travel only makes sense with an understanding of the potential scale of space manufacturing activities, opportunities for space manufacturing will be reviewed before turning to interstellar settlement.

Consider first the immediate future. Manned activities in space on a regular basis begin with the space shuttle, scheduled to start operation around the time of publication of this article. Unless development capital is significantly curtailed, it is reasonably likely that construction of a small permanent manned orbital laboratory will begin in this century to support existing activities such as reconnaisance, communications, and research.

It is instructive to try to estimate the cost of these space activities in units which relate directly to the human effort involved, in order that factors such as inflation and changing types and efficiency of manufacturing processes will not fundamentally alter the intuitive meaning of the units used. A useful unit is "millions of person-centuries" (mpc), where one person-century is 200,000 hours of human labor. Roughly 0.1 mpc have been expended on space activities over the thirty years leading to deployment of the space shuttle. Given the importance currently attached to the military, economic, and scientific activities discussed above, and the capital invested and the relative efficiency of the shuttle as a launcher, this rate of expenditure is unlikely to decrease by a large factor in the forseeable future.

Discussion of possibilities for further human activities in space centers on industrial processes such as solar power satellite stations, small scale production of special materials or biological products, and large scale processing of lunar or asteroidal resources. An important factor concerning human activity in space is that, in the long term, there is considerable motivation to turn to use of extraterrestrial sources for materials to build life support systems for the people involved in space operations with continued human interaction. The motivation is the potential for orders of magnitude reduction in transportation costs for acquiring materials from the moon (or Earth-approaching asteroids) compared to launching from the surface of the Earth. For very extended or large scale space activities, the impact on Earth resources and environment could also be minimized by restricting most transportation and manufacturing to outside the Earth's atmosphere.

It is therefore appropriate that a revival of interest in constructing habitats in space followed the suggestion by O'Neill and coworkers that bags of soil could be launched from the moon and used to construct a self-sufficient ecosystem at a "moderate" cost. Since elaborations of this proposal have been extensively described elsewhere, only features particularly relevant to the problems of interstellar settlement will be reviewed here. Topics of particular interest are propulsion, ecosystem design (including mass per person, radiation shielding, leakage rates, and minimum size), cost, and schedule.

The main propulsion requirement in O'Neill's proposal was for moving the main mass of structural material for the space habitat to its construction site in high Earth orbit. The solution was to place a linear synchronous motor (electromagnetic mass driver) at an appropriate elevated point on the lunar equator. Bags of lunar soil launched from the mass driver at 2.4 km/s would have to be very accurately aimed to reach a catcher device placed beyond the moon. This could be accomplished by insuring that the payload goes through a small aperture about 150 km downrange, across a lunar valley. This accuracy is thought to be readily achievable with a set of progressively more distant course correction devices which charge the payload, measure its position, deflect it electrostatically, and then discharge it. In fact, with a 150 km baseline for course correction, it should be possible to launch the payload much more accurately than one can compute the solar and lunar and other perturbations on its orbit. Since these perturbations are very small, it would be possible to use a catcher only a few meters across (cf. articles in O'Neill and O'Leary 1977).

The first comprehensive ecosystem design for a habitat for ten thousand people (Johnson and Holbrow 1977) used seventy-four tonne/person, the majority of which was for soil and for the container. An additional kilotonne per person of industrial slag was allocated for radiation shielding, largely to protect against solar cosmic rays. Various aspects of the ecosystem recycling problem were analyzed, and it was concluded that it should be possible to design a system with 100 percent recycling which could retain its atmosphere for centuries. The minimum size of a closed human ecosystem (cf. Gilligan 1975) was not directly addressed in these studies, because the space habitats were sized for the manufacturing capability. In particular, about ten thousand inhabitants were required in a space manufacturing facility so that in ten years they could build several 5 GW power satellites or build the structure for a second habitat for ten thousand people. It seems likely from the results of these studies that it would be <u>possible</u> to build a smaller <u>self-contained</u> ecosystem with 50 to 500 people with 10,000 to 100,000 tonnes of structure and biological mass. (It might be cheaper to supplement such a habitat with materials from Earth in the <u>initial</u> stages of space settlement, but we are concerned here with a different situation where it is only necessary to scale down existing larger ecosystems.)

The cost of building a manufacturing facility with ten thousand occupants in twenty-two years has been calculated to be 2×10^{11} 1975 United States dollars, or about 0.2 million person-centuries. While this may be optimistic, the cost estimate is sufficiently detailed so that it is unlikely to be off by an order of magnitude, which is sufficient for the present discussion.

A further development in space manufacturing would be retrieval of small Earth-approaching asteroids (Arnold and Duke 1978, O'Leary 1978). Some of these contain a better mix of materials for building ecosystems than lunar soil does. Others could serve as a vast source of metals for space or even terrestrial manufacturing. Estimates of the cost of transporting the best candidates to Earth orbit are already competitive with the lunar alternative. With a better survey of some of the estimated 100,000 Earth-approaching asteroids of appropriate size, and with possible application of very efficient retrieval methods (Singer 1979), such asteroids may become the initial as well as eventual source of materials for large space habitats.

The rapid construction proposed in the studies described above was timed to provide large supplies of electricity from solar power satellites in the near future. Whether this comes to pass will be determined by the success of pilot projects to be deployed in the next few years. Otherwise, the habitation of space will probably procede at a slower pace determined initially by servicing of reconnaissance, communications, research, and (Driggers 1977) specialized manufacturing processes. Eventually, the lower average cost of sustaining such activities should dictate use of closed ecosystems supplemented with extraterrestrial resources, but this could take many decades. A key factor in the habitation of space is the production of one space manufacturing facility capable of building most of a second such facility, but it is difficult to predict whether this will take as little as twenty or as much as two hundred years.

To complete the discussion of settlement of the solar system, we note that designs of space settlements either for one hundred thousand or for over a million people have been outlined (e.g. O'Neill 1974, and references quoted above). While the living conditions in a smaller space habitat are meant to be comparable to those in an affluent city or existing small village, these larger habitats may include open spaces with a scale of many kilometers and possibly even weather. Whether large or small habitats are chosen, it seems clear that the possibility exists to have millions of people living in many structures in space. In the author's opinion, this is remotely possible within something over fifty years, probable within five hundred years, and difficult to avoid within less than five thousand years (barring profound biological transformation, incessantly repeated global catastrophes, or the remarkably difficult (Viewing and Horswell 1978) achievement of catastrophe leading to human extinction).

"KNOWN" SOLUTION TO BIOLOGICAL PROBLEMS
OF INTERSTELLAR SETTLEMENT

Since interstellar transport would very likely require large construction activities in space, the above discussion has concentrated on space settlement, and ignored additional possibilities concerning settlement of planets and satellites. (These intriguing additional possibilities are discussed by J. Oberg at this conference; they too would probably require substantial habitation in space settlements as a prerequisite.) The discussion will now concentrate on what would be required for some of the inhabitants of the solar system to move to a nearby star. First the biological problems of the spaceship design and then the propulsion problem will be discussed.

Staying within the context of the present discussion, the only known biological solution is to use a pre-existing closed ecosystem as the interstellar spaceship (i.e., an "interstellar ark," Matloff 1976). The fundamental environmental design problems for this solution have been extensively addressed in the space habitation studies, as discussed above. Other possible solutions are too speculative to help give any confidence in estimates of maximum requirements for interstellar settlement, so discussion of these is relegated to Appendix A. Moreover, the technological base required for presently known solutions to the propulsion problem cited below already requires extensive experience at designing space ecosystems, and inhabiting them for at least a century or more. Therefore, discussion is restricted to a few biological problems which could be unique to this method of interstellar settlement.

One suggested biological problem is in breeding due to a small genetic pool. This is clearly a "red herring" for two reasons. Firstly, proper crew selection should introduce great genetic diversity (making the dubious assumption that this is indeed essential). Secondly, the already widespread practice of storing and transferring human zygotes could provide unlimited diversity with a few grams of such cells.

Another problem is radiation exposure from galactic cosmic rays. In the space habitat studies it was necessary to provide significant additional shielding mass, partly for protection against some galactic cosmic rays, but primarily for protection against solar cosmic rays. Since active shielding against such radiation cannot necessarily be provided during the interstellar voyage, one solution is to accept the order of magnitude increase in payload mass required by passive shielding. Another possibility is to accept the 5 to 20 rem/yr (depending on detailed design) dose without massive shielding (compared to a minimum of 50 rem/yr quoted as a lower limit required for detectable damage from steady radiation, Johnson and Holbrow 1977). Crew selection might avoid rare (but perhaps presently epidemiologically significant) radiation sensitive individuals. Extra shielding for zygotes, and possibly for developing organisms, might also be useful. Since the possible hazards from failure of energy supply (e.g. a several megawatt nuclear reactor), environmental systems, and especially propulsion systems are likely to greatly outweigh the radiation hazard, it is likely that multiple redundancy in these systems would enhance the probability of success much more than passive shielding.

Finally, there is the question of social stability in a small remote habitat. The following observations may be relevant to this question. First, there is no known universal tendency for small isolated human communities to inevitably self destruct. Second, the possibility of continuous habitation of space habitats in the solar system for centuries could allow extensive opportunities to test social stability under related physical conditions. Third, the most promising propulsion method discussed below could be facilitated by extended human support in the local interstellar medium. This would allow a further testbed for social stability in interstellar arks, should it be necessary. All of the biological and social problems mentioned above look relatively tractable in comparison to the propulsion requirements, which are now discussed.

AN EXAMPLE OF PROPULSION FOR HUMAN
INTERSTELLAR SETTLEMENT

Until recently, the best interstellar propulsion method for which reasonable confidence of feasibility can be claimed was the pulsed thermonuclear bomb and pusher plate concept outlined by Dyson (1968). Two other concepts appear interesting but have potentially unsolvable engineering difficulties. These are microfusion explosions ignited by relativistic electron beams (Bond et al. 1978), and use of a highly collimated laser beam pointing out from the solar system (Marx 1966, Jackson and Whitmire 1978). The microfusion concept is plagued by uncertainties that plasma stability will allow a net energy gain (much less the high efficiency required for interstellar propulsion). Another problem is the damage potential of the enormous number of neutrons produced in any reasonable microfusion scheme; it is quite likely that detailed results will show that it is not possible to guarantee re-absorption of nearly all of these neutrons during the complete sequence of an efficient microexplosion. While beaming laser energy to an interstellar spaceship would reduce onboard propulsion requirements dramatically, it is far from clear that the required laser optics (e.g. sources about 100 km square with a tolerance below one part in 10^{11}, Singer 1980) are achievable. (The requirements for using laser power for deceleration are even more formidable, although alternative solutions for deceleration may exist, cf. F. Dyson's presentation at this conference.)

Fortunately, it has recently been possible to look in some detail at another propulsion method which combines the advantages of those discussed above; these are off-board propulsion, relatively high efficiency, and reasonable confidence of engineering feasibility. This propulsion method uses an electromagnetic mass driver to launch a stream of small pellets using a local solar or nuclear power source (Singer 1980). An analysis of this method was stimulated by work on mass drivers at Princeton (e.g. O'Neill and Snow 1979), and MIT (e.g. Kolm et. al. 1979), and more recently at the Lawrence Livermore Laboratory (Brittingham and Hawke 1979), by the achievement of acceleration of 360,000 gravities on a 5 meter rail gun (Rasleigh and Marshall 1978), and by investigation of interstellar propulsion problems carried out by the Daedalus team (Bond et al. 1978). It was also preceeded by proposals for hypervelocity accelerators by Winterberg (1966), for local applications by Ruppe (1966), and on the potential of mass drivers by Clarke (1950). Another critical observation was the suggestion that projectiles could be aimed extremely accurately by successive course corrections over ever longer baselines, as discussed above for the moon based launcher. Relatively pessimistic assumptions about interstellar dispersion of a collimated pellet stream show that a modest number of course correction devices would be required to retain collimation of the stream over distances of lightyears. Finally, the pusher plate concepts used in the thermonuclear bomb and microfusion schemes were readily adapted to the pellet-stream concept, with the advantage that ionization of the pellets near the pusher plate would give a clean low-temperature plasma with negligible neutron flux.

One of many possible scenarios using pellet-stream propulsion has been analyzed to provide a concrete example of an interstellar settlement mission. The sequence of events in this mission is illustrated schematically in figure 7.1. The events are as follows:

1. Preparation for this mission begins 110 years before the departure of the settlers with the launch of two slow pellet streams from a relatively short mass driver. (The mass driver is shown with its length outlined against the sun in the upper left of figure 7.1.) These slow pellet streams will be used for partial deceleration of the payloads before they reach the target, Proxima Centauri. The upper, faster stream will be used by a small lead ship and the lower, slower stream by the main body of settlers.

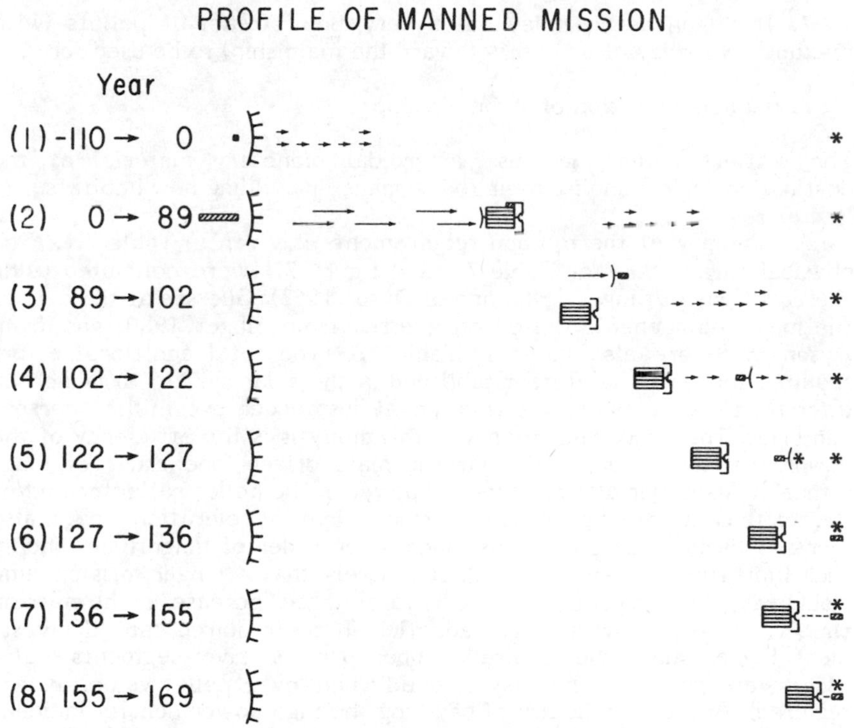

PROFILE OF MANNED MISSION

Year

(1) -110 → 0

(2) 0 → 89

(3) 89 → 102

(4) 102 → 122

(5) 122 → 127

(6) 127 → 136

(7) 136 → 155

(8) 155 → 169

Fig. 7.1. Schematic illustrations of steps in an interstellar settlement mission. Arrows on pellets indicate velocities of the various pellet streams used for acceleration and partial deceleration.

2. For 89 years, faster pellets travelling 39,000 km/s are intercepted by the main ship. The pellets are launched from a mass driver consisting of 100,000 segments each 3,000 meters long.

3. For another 13 years, another stream of similar pellets continues to accelerate the lead ship.

4. First the lead ship and then the main ship partly decelerate by running into the slow pellet streams.

5. The lead ship eventually stops using fusion bombs and a pusher plate for deceleration (an alternative deceleration method is outlined in Appendix B). Thirty-eight years after separating from the main ship, the lead ship arrives at Proxima Centauri and,

6. begins construction of a third miniature pellet launcher.

7. This launcher will lay out a very slow stream of pellets (with maximum velocity of 610 km/s toward the main ship) to be used for

8. final deceleration of the main ship.

The settlers could then use asteroidal planetary material at the destination for expanding their living space or building new habitats.

A summary of the mission requirements is given in Table 7.1. (The physical parameters for Table 7.1 and figure 7.1 were computed using the equations of Singer 1980, and of Dyson 1968). Guesses at the cost of the major components (based on guesses from Singer 1980, and from Dyson 1968) are also listed in Table 7.1. The total additional effort required for space habitats established in the solar system to launch an interstellar settlement is estimated at just under one million person-centuries. The major uncertainty in this analysis is the efficiency of the downrange segments of the largest mass driver (one order of magnitude?). A smaller allowed specific power in the pellet reflector and/or the efficiency of the pulsed thermonuclear deceleration could also increase the mission cost by as much as an order of magnitude. Should such limitations be encountered, it is likely that a longer mission time would be chosen. Even a factor of two or three increase in the mission time t_* would allow a large reduction in the required capital investment. For example, the required number of mass driver segments scales as t_*^{-2} (and they may be easy to build when lower pellet velocities are required). And the difficulty of handling the high power density incident on the spaceship's reflector plate decreases at least as t_*^{-3}. Thus it seems likely that settlement missions requiring 170 to 500 years of travel time could be achieved with a total investment within an order of magnitude of one million person-centuries.

Table 7.1. Nominal Mission Requirements
with "Present" Technology

Component	Size		Cost (mpc)
Launcher			
(a) accel	10^5 of 3000 m segments		0.3
(b) decel	2000 of 3000 m segments		0.01
Power supply	40 TW maximum		0.1
	Main Vessel	Lead Ship	
Ship Mass	100 ktonne	100 ktonne*	0.01
Thermonuclear Fuel	———	1000 ktonne*	0.3
Settlers	1000	100	0.002
		Total =	0.72

*Most of the mass of the lead ship and its thermonuclear fuel is delivered in another pellet stream (not shown in Fig. 7.1). The velocity of the lead ship after acceleration is adjusted to approximately match the velocity of this supply stream.

CONCLUSIONS

The above discussion can be summarized very succinctly as follows. We can be reasonably confident that it should be physically possible to begin to send settlements to nearby stellar systems after establishing a population of many millions in space habitats in the solar system. This is likely to happen several centuries after the invention of the radio telescope, and could conceivably require only one century. The time required to set up such habitats with a modest fraction of the available manpower is in any case certainly much less than 10^4 years. A maximum limit of two to five centuries on the time required to transport a self-contained human ecosystem to the nearest stars can be inferred from our present understanding of the biological and physical problems involved. Establishing the capability to launch such missions should be possible with an investment on the order of a million person-centuries.

Now consider the implications of these conclusions for the argument that interstellar settlement is physically unrealistic for cultures which are potential sources of communication from ETI. Since the technological lifetimes of these cultures exceeds 10^4 years in even the most extreme scenarios supporting ETI search strategies, it appears that every source of communication from ETI in the galaxy has had ample opportunity to instigate interstellar settlement. Whether any of them would want to has been questioned. The only comment made here is the following. If ETI have resources comparable to our own, then success of an ambitious ETI search would imply that the equivalent of at least 10^{15} mpc had existed in the galaxy prior to our receiving the first communication. The comparison of this number to the order of 1 mpc required for us to initiate interstellar settlement makes an impressive contrast.

Although a careful analysis of the full implications of these results is beyond the scope of this presentation, one possible conclusion is evident, that it may be extremely difficult or impossible to find sources of communication from ETI in the galaxy. In fact, in the light of a sober reappraisal of our minute but increasing understanding of the large variety of possible factors in astophysical and chemical dynamics which may be required for the appearance of technological intelligence, it would be far from surprising if we were the only technological culture in the galaxy. In this case, in the words of a reporter who covered this conference, "perhaps it is the great destiny of man himself to spread life through the galaxy."

Hopefully, it will be obvious that the important question is not how quickly we can fulfill such a destiny, but rather of what the quality of the life we spread through the galaxy will be. These may be the sentiments expressed in this fragment of a poem by Richard Brautigan (1977):

I like this planet.
It's my home and I think it needs our attention and our love.
Let the stars wait a little while longer.
They are good at it.
We'll join them soon enough.
We'll be there.

APPENDIX A. ALTERNATE MODES OF
BIOLOGICAL TRANSPORT

The discussion of interstellar settlement in the main text was restricted to consideration of small human communities reproducing by methods already considered acceptable. This assumption is useful for studying the near future or for constructing an "existence proof" of the possibility of interstellar settlement. But it is far from clear that it is safe to assume the same biological problems will be faced by ETI or even by our own progeny in the distant future. Given that recent natural biological evolution has produced profound changes in the form of intelligent life on Earth in a minute fraction of galactic history, there would certainly be further evolution even without human intervention.

However, it is becoming increasingly clear that over the next few decades and centuries, we will acquire increasing control over technology relevant to genetics. At the least, this will probably involve decoding the structure and function of a significant part of human and other genomes, and the ability to construct genes of any desired nucleotide sequence and insert them into many kinds of cells. That there will be a concerted attempt to use such technology to effect certain improvements in human genomes (e.g. cure of genetic defects and/or of various diseases) is also already quite clear. It seems plausible to suggest that the range of possible further developments in genetic engineering allowed by the technological base will be so large that social rather than technological factors will limit what is actually done.

The range of biological possibilities in a galaxy with a large number of long-lived technological ETI (as presumed in most communication search strategies) is evidently enormous. To begin with, it is only a plausible but unprovable hypothesis that natural evolution of ETI primarily occurs on timescales and in physical conditions which produce (to within a modest factor) ETI with size and lifetime comparable to ours. Were there large numbers of ETI in the galaxy, then the possibility of artificial control of genetics to produce mature ETI of different size, and particularly different longevity, might well be available to many cultures or subcultures. Moreover, one can imagine different modes of reproduction which produce physically similar mature ETI but which could have profound implications for interstellar settlement.

Only a few of these myriad possibilities will be discussed. In particular a discussion of the traditional example of hibernation of human communities is avoided. Hibernation would still require some life support system during transit and might very likely require a sizeable ecosystem after arrival, at least for travel to nearby stars. It might therefore only save a modest factor in the mass to be transported. Moreover, since primates do not hibernate it might also require significant genetic redesign. Therefore, only more interesting examples of the possibilities with active control of genetics are discussed here. It is certainly not claimed that we or other ETI could or would necessarily want to achieve all of these possibilities. But it is claimed that the existence of something similar to at least some of these alternatives is plausible. The range of biological alternatives make it extremely difficult to place any lower physical limit on the speed of interstellar settlement which is relevant on the timescale of galactic evolution.

First, consider ETI which have achieved longevity and psychological stability of 1,000 years of more. One or two such individuals could conceivably be sent on a settlement journey in a ship orders of magnitude smaller than the large interstellar ark described in the main text. Particularly if large bombs were not necessary for deceleration (cf. Appendix B), this could allow very small spaceships. The ability to undertake millenial voyages could also greatly relax the propulsion requirements. If the individuals were, say, an order of magnitude smaller than humans, it could reduce the propulsion requirements still further. If necessary, they could presumably carry along a sperm or zygote bank. (Building a radio receiver to receive the latest word on desirable nucleotide sequences for genomes might be an alternative to a zygote bank.)

Alternatively, the zygotes could even be matured at destination by an automatic device, assuming this would result in lower spaceship mass than carrying a mature ETI. This may seem a highly unpalatable prospect from a human point of view, but it cannot be logically excluded.

Another possibility with some terrestrial analogues is that an ETI might have naturally or intentionally adapted to living in space in slightly more radical ways. A few millimeters of protective film exuded on the surface of an ETI (or a nonsentient biological "house") might allow direct implantation and growth of a small immature ETI on the surface of an asteroid in space. Should deceleration methods like those described in Appendix B be possible, then an immature ETI with a mass of a gram or less might be launched directly from a long mass driver with relatively modest acceleration in a "ship" of mass from a gram to a tonne. Contrary to current ideas, it could therefore theoretically be possible for a single ETI to send self-reproducing "settlements" at very high velocity to millions or billions of stellar systems, even to neighboring galaxies.

This last suggestion goes far beyond what is necessary to demonstrate the importance of considering biological alternatives when attempting to make definitive statements about the limits to interstellar settlement, and it may in fact be physically impossible. But the less radical examples clearly demonstrate that assuming known biological alternatives is only useful in putting an _upper_ limit on the difficulty of interstellar settlement.

APPENDIX B. DECELERATION ON BLOWBACK FROM COLLIDING PELLETS

Deceleration is a major problem for a stellar orbiter probe or settlement mission using any propulsion method. For onboard propulsion, either the mission time is doubled or the fuel requirements are typically greatly increased in comparison to a fly by mission. Were remote electromagnetic propulsion made practical (e.g. by periodic refocusing of the beam) deceleration would be particularly problematic. Deceleration on a slow pellet stream is promising, but it is limited by velocity of the prelaunched pellets which the decelerating ship runs into.

The alternative of electromagnetic braking on a set of fine wires (with minimum size limited by erosion) was discussed by F. Dyson at this conference. Assuming a localized hyperalfvenic shock forms around each wire, there is a question as to whether the resulting turbulent wake would produce significant drag. (A reliable answer to this question would require experimental support.)

(a) Before Pellet Impact

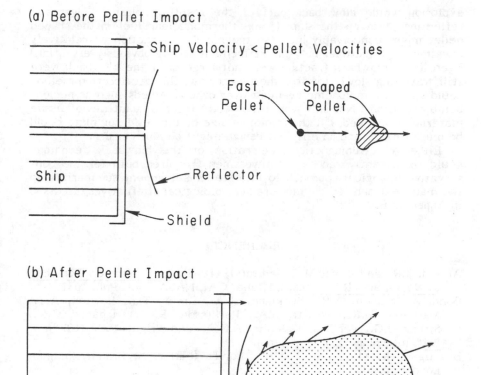

(b) After Pellet Impact

Fig. 7.2. Schematic illustration of blowback deceleration. Arrows indicate relative velocities as viewed from Earth. Sizes of ship and pellets are not to scale.

An alternative suggestion arising as a result of this conference is to produce a stationary plasma in front of the decelerating spaceship using colliding pellets. This idea is illustrated in figure 7.2. Two streams of pellets launched well before the beginning of the mission would be collimated as described in the text. Longitudinal velocity corrections by device(s) released from the spaceship would insure that pairs of pellets from each stream pass through a collimator on the ship near the same time. A slower, shaped, pellet would be overtaken by a smaller faster pellet. The slower pellet would be shaped so that the resulting impact explosion would blow back part of the resulting plasma towards a reflecting plate on the ship. (Some thermonuclear fuel in the shaped pellet might conceivably increase the blowback fraction at relatively low impact velocities, but this refinement is not obviously essential.) Even if the blowback fraction were not large or if the blowback were still travelling slowly towards the target star, this deceleration method could still be a major improvement over other methods such as nuclear bomb explosions. Not only would the need for large quantities of fissile material be avoided, but the minimum size of the reflector plate could be much smaller and an efficient design might be much easier.

Either electromagnetic deceleration or the blowback technique could significantly reduce the investment required both for the conservative biological approach to interstellar settlement summarized in the main text and for the more radical biological alternatives discussed in Appendix B.

REFERENCES

Arnold, J.R. and Duke, M.G. (editors), (1978). 1977 Summer Workshop on Near-Earth Resources, La Jolla, CA., NASA Conf. Pub. 2031.

Bond, A., Martin, A.R., Buckland, R.A., Grant, T.J., Lawton, A.T., Mattison, H.R., Parfitt, J.A., Parkinson, R.C., Richards, G.R., Strong, J.G., Webb, G.M., White, A.G.A., and Wright, P.P., (1978). J. Brit. Interplanetary Soc. Suppl.

Brautigan, R. (1977). In Space Colonies (editor: S. Brand), Penguin London, p. 51.

Brittingham, J.N., and Hawke, R.S. (1979). Devices for Launching 0.1-g Projectiles to 150 km/s or More to Initiate Fusion, UCRL-52778.

Clarke, A.C. (1950). J. Brit. Interplanetary Soc., 9, 261.

Cocconi, G., and Morrison, P. (1959). Nature 184, 844.

Driggers, G.W. (1979). Is Lunar Material Use Practical in a Non-SPS Scenario? Paper 79-1414, Fourth Princeton/AIAA Conference on Space Manufacturing, Princeton.

Dyson, F.J. (1968). Physics Today 19, (Number 10), 41.

Gilligan, E.S. (1975). Migration to the Stars (Luce, Washington).

Grey, J.G. (editor), (1977). Space Manufacturing Facilities, Vol. 1 and Vol. 2 (Proc. 1974, 1975, and 1977 Princeton/AIAA Conference on Space Manufacturing Facilities), AIAA: New York.

Hart, M.H. (1975). Quarterly J. Royal Astron. Soc. 16, 128.

Jackson, IV, A.A., and Whitmire, D.P. (1978). J. Brit. Interplanetary Soc. 31, 335.

Johnson, R.D., and Holbrow, C. (editors), (1977). Space Settlements: A Design Study, 1975 Stanford University-AMCS Research Center Summer Faculty Fellowship Program in Engineering Systems Design, NASA SP-413.

Jones, E.M. (1976). Icarus 28, 421.

Kolm, H., Fine, K., Mongeau, P., and Williams, F. (1979). Electromagnetic Propulsion, Paper 79-1400, Fourth Princeton/AIAA Conference on Space Manufacturing Facilities, Princeton.

Marx, G., (1966). Nature 211, 22.

Marx, G., (1973). In Communication with Extraterrestrial Intelligence, (editor: C. Sagan), MIT Press, Cambridge, Mass. p. 226.

Matloff, G.L., (1976). J. Brit. Interplanetary Soc. 29, 775.

Michaud, M.A.G., (1979). J. Brit. Interplanetary Soc. 32, 116.

Oliver, B.M., and Billingham, J., (1971). Project Cyclops: A Design Study of a System for Detecting Extraterrestrial Life, NASA CR 114445.

O'Leary, B., (1978). Astronomy (number 6), p. 6.

O'Neill, G.K., (1974). Physics Today 27, (number 9), 32.

O'Neill, G.K., and O'Leary, B. (editors), (1977). Progress in Astronautics and Aeronautics, 57, Space-Based Manufacturing from Non-terrestrial Materials; 1976 NASA-Ames Study, AIAA: New York.

O'Neill, G.K., and Snow, W.R., (1979). Overview and Outline of Mass-Driver Two, Paper 79-1396, Fourth Princeton/AIAA Conference on Space Manufacturing Facilities, Princeton.

Purcell, E.M., (1960). In Interstellar Communication, (editor: A.G.W. Cameron), W.A. Benjamin, New York, p. 121.

Rasleigh, S.C., and Marshall, R.A., (1978). J. Appl. Phys. 49, 2540.

Ruppe, H.O., (1966). Introduction to Astronautics and Aeronautics Vol. 1. (Academic, New York), p. 20.

Singer, C.E., (1979). Collisional Orbital Change of Asteroidal Materials, Paper 79-1434, Fourth Princeton/AIAA Conference on Space Manufacturing Facilities, Princeton.

Singer, C.E., (1980). J. Brit. Interplanetary Soc. 33, 107.

Viewing, D.R.J., and Horswell, C.J. (1978). J. Brit. Interplanetary Soc. 31, 209.

Winterberg, F. (1966). Plasma Phys. 8, 541.

Winterberg, F. (1979). J. Brit. Interplanetary Soc. 32, 403.

8
Terraforming
James Oberg

Earth is unique in this solar system — it is the only planet that seems to support life. Its hospitable ecosphere stands in stark contrast to the empty, lifeless landscapes of the moon, Mars, Venus, Mercury, and other worlds probed by our spacecraft.

But it doesn't have to stay that way. People may some day be able to increase the number of life-supporting worlds in this solar system from one, as at present, to a dozen or more. Instead of just being a freak accident of biology, Earth could serve as a blueprint for the transformation of her sterile sister worlds into earthlike biospheres. There's no need to wait for this metamorphosis to come about through billions of years of random accidents. It can be made to happen in just a few centuries of deliberate human manipulation.

The term for this awesome concept is terraforming. The word was coined 40 years ago in a science-fiction story, but the concept of world shaping goes back much further. How far it may still be ahead of its time is unknown.

Advocates of terraforming conjure up idyllic visions of reshaped planets made fit for human settlements. They ask incredulous listeners to imagine red-skyed Mars watered and in bloom, to imagine choking Venus tamed and cooled, to imagine even the sterile, airless moon transformed into a smaller replica of Earth.

And how could such remodeling be accomplished? The transformations, it turns out, do not really require magic but only extrapolation from what we know today. Colossal energies would be needed, but in a few centuries such forces should be available. Intimate knowledge of climatology would be needed, but those lessons must be learned on earth in any case. Manipulations of biology and ecology are still far beyond present scientific capabilities, but they are in the directions along which modern science is moving.

Each planet would require a different combination of these techniques, although many of the tools would be common from world to world. The major obstacle is not technological but conceptual: Humanity does not yet realize that it has the capability to transform whole planets, for worse (as we are often warned by the doomsday prophets) and for better.

A good first candidate is the planet Venus, once thought to be a twin of the earth but now known to be a closer analogue of medieval visions of hell. The planet gets too much sunlight, has too much carbon dioxide and sulfuric acid, and has a day-night cycle far too long.

Imagine what would happen to a human being placed on the surface of Venus. Immense atmospheric pressures would instantly crush the soft body tissues, while ovenlike temperatures would convert body water to steam.   Layers of charring flesh would peel off explosively. An expanding cloud of soot would surround a pile of crumbling bones as the acidic vapors of Venus turned a human body to dust.

Terraformers call for a physical assault on such hostile conditions. Carbon dioxide in the atmosphere of Venus would be transformed biologically by clouds of algae, suitably tailored in genetics laboratories to thrive under Venusian conditions. Artificial dust clouds would shade the planet. The excess oxygen in the atmosphere (about sixty bars worth) would have to be physically ejected or – better yet – combined with hydrogen imported from the atmosphere of Saturn, to form water oceans.

A century or two after this transformation, human beings would walk in the upland regions of Venus without backpack refrigerators and perhaps even find it pleasant. Along the newly formed coasts of the Aphrodite and Lakhshmi continents, the climate could resemble that of Samoa or Curacao or the Cote d'Azur.

Mars offers different problems. A human body placed on Mars would exhale all of its internal gases in a great rush out of body orifices. Consciousness would fade through anoxia as the thin Martian air provided no breath. The cold of the sand would freeze the fallen body within minutes, but it might take millions of years for the mummified flesh to erode. In the end, one patch of red sand might have a lighter coloration; beyond that, the visitor would leave no trace.

Mars has neither enough air nor enough sunlight, although its day is nearly earthlike. To attract more sunlight, dark soot could be mined from the two small moonlets in Mars's orbit and spread on the surface of the planet. If permafrost (giant dust-covered glaciers) exist, it might melt, flooding the surface after a billion years of drought; in cases where such planetary water does not exist, asteroidal water would have to be imported. Biological activities could be instigated, perhaps in miles-deep oasis valleys gouged out of the landscape by the ice melt or by the impact of incoming asteroids. Additional heating could be provided by giant space mirrors, a thousand kilometers on a side, concentrating sunlight onto the planet.

As the Martian air thickens to breathable levels, and as temperatures rise to above the freezing point of water, a new climate could be formed that would approximate that of the Andes or the Caucasus or Kashmir. If such conditions on the new planet seem unattractive, recall the persistent stories of longevity and happiness among mountain people here on earth.

Earth's own moon need not remain forever barren. Even as human mining activities bring it a measure of life in the coming decades, it too could hold an atmosphere, either baked from its own rocks or imported from Saturn and beyond. Other rocky worlds such as Mercury and various moons of the outer planets such as Ganymede, Titan, and Triton could similarly be engineered into habitable home worlds.

There is a new candidate whose qualifications have only recently been recognized: Io, innermost of the Galilean worlds circling Jupiter. Its sulfur volcanoes blasted their way into human consciousness via the Voyager probes, driving home the lesson that in space one can only dare expect the unexpected. It is in keeping with this still-unfaded sense of wonder, elicited by the Voyager vistas, that I want to nominate Io for terraforming – and at the top of the heap, as well.

Io has several advantages over more classical candidates. First, it has an internal heat source generated by tidal stresses induced by Jupiter (most of the surface is cold because in a vacuum the outflowing heat leaks away quickly). Second, its 42 1/2-hour day is not grossly different from that of Earth. Third, it is deep within the magnetic field of Jupiter, a factor which has biochemical advantages lacking on all other solid planets beyond Earth.

But of course Io has some powerful drawbacks, at present. First is the killing radiation belt which surrounds Jupiter. Second is the lack of water or, for that matter, any atmosphere worth sneezing into. Third is the surface enrichment of sulfur compounds spewed forth from underground lakes of molten sulfur. Each of these problems by itself might seem to veto any consideration of terraforming Io and appear to counterbalance the substantial advantages enumerated above.

Well, maybe not. Radiation belts are swept out by rings of rocky debris – as we learned most recently and dramatically during the Saturn flyby. The Jovian radiation in the neighborhood of Io could thus be decontaminated by pulverizing a large fraction of the small inner carbonaceous moon, Amalthea (or the even smaller moons discovered inside Amalthea's orbit), forming an artificial ring extending out to and enveloping Io.

Next, water would have to be imported, either from ice-rich, sibling-worlds Ganymede and Callisto or by impacting some outer Jovian moons or nearby asteroids onto Io (comets are too small and too unpredictable). Note that the sulfur, while plentiful on the surface due to differentiation, should not be more abundant in proportion to the whole mass of the planet than it is on Earth. The surface sulfur could be buried using dirt excavated by the impact force of the "ice-teroids" carrying the components of the future atmosphere and ocean. Subsequently, biological tailoring could begin, lasting many decades – then spacesuits on Io would become obsolete. Io could be made a habitable world by the end of the next century.

Or maybe not. Perhaps its thermal state is too active to support a stable crust (then we can turn our attention to cooler, wetter Europa).

Perhaps the asteroidal engineering needed to "short circuit" the Jovian radiation belts is too ambitious or unreliable. Perhaps some form of carbon-sulfur life exists in the hot springs or, even more bizarre, swims in the liquid sulfur ocean beneath the crust.

Just imagining the environmental impact statement for such a project is enough to make my hair stand on end.

Terraforming studies consider the conditions needed on habitable planets, and the tools and techniques which are conceivable today. These topics fill several chapters in my forthcoming book, New Earths (Stackpole, 1981).

Climatological goals can be quantified; roadblocks can be identified and areas of current ignorance defined; strategies can be suggested. The bottom line is that terraforming is possible given enough time and enough money and enough intelligence.

If Earth-like worlds are truly rare in the universe, and if star-faring civilizations retain a desire to live on (or even just vacation on) planetary surfaces, they may not be able to find sufficient hospitable planets. In that case they may have to make their own – and everything so far indicates that such planetary engineering is feasible.

9
Estimates of Expansion Time Scales
Eric M. Jones

ABSTRACT

Monte Carlo simulations of the expansion of a spacefaring civilization show that descendants of that civilization should be found near virtually every useful star in the Galaxy in a time much less than the current age of the Galaxy. Only extreme assumptions about local population growth rates, emigration rates, or ship ranges can slow or halt an expansion. The apparent absence of extraterrestrials from the solar system suggests that no such civilization has arisen in the Galaxy.

INTRODUCTION

"I see that the valleys are thick with people and even the uplands are becoming crowded. I have selected a star, and beneath that star there is a land that will provide us with a peaceful home."

- Ru, Traditional Founder of Aitutaki in the Cook islands (Buck 1938)

An important part of the question, "Where are they?" is the question "Could they have gotten here yet?" If we imagine a spacefaring civilization arisen a billion years ago and a thousand parsecs from Earth, what are the odds that the descendants of that civilization would have established settlements in the solar system before now? The answer, I believe, is that, if such a civilization had arisen and if interstellar travel is practical at a few percent of light speed, it is virtually certain that the solar system would have been settled by non-natives long ago. Unless we discover that interstellar travel is impractical, I conclude that we are probably alone in the Galaxy.

We know nothing of any extraterrestrial civilization. If we assume some have existed, it is also reasonable to assume that at least some would be as inquisitive and as eager for adventure as humanity (Hart 1975). It would take but one such species to fill the Galaxy.

Humanity has a history of expansion into available areas on Earth. If we examine our past we can estimate how long it might be before humanity would expand throughout the Galaxy. If that time is greater than a few billion years, we would conclude that the question "Where are they?" isn't very meaningful. But if the time is less than a billion years or so, the apparent absence of extraterrestrials from the solar system is significant.

I estimate that if we develop the technical means for practical interstellar travel, humanity and its descendants will fill the Galaxy in a relatively brief time — 300 million years at most, the most likely value being 60 million years. This estimate is based on assumptions about local rates of population growth, the rate at which emigrants leave one place for another, and on the choice of a mathematical model. The choices are debatable. Newman and Sagan (1978), in a widely distributed preprint, have argued for very low rates and for a different mathematical model. They estimate that the expansion time — the time to fill the Galaxy — exceeds 10^9 years. Let us examine the expansion/settlement process and the choice of rates. Afterwards, we will briefly discuss the mathematical models and, finally, discuss estimates for the expansion time.

HUMAN EXPANSION

Our ancestors seem to have begun in East Africa one to two million years ago. From earliest times humans have lived in small gatherer/hunter bands which diffused throughout the Old World (Africa, Europe, and Asia). The dispersal across the Old World was undoubtedly unplanned. A band might chance to move a little further from East Africa because of movement of game, local climate changes, population pressures, warfare, disease, or natural disaster. The decision to move was made within the band (perhaps not even consciously), but the net result of millions of such decisions was that our ancestors, members of a clever and adaptable species, inhabited all parts of the Earth that were physically accessible to them before the rise of agriculture some ten thousand years ago (Davis, 1974). Although humanity spread across the face of the Earth, the population increased only slowly. The rate of increase during humanity's gatherer/hunter existence has been estimated to have been only .0015 percent per year (Coale 1974). The available technologies could support only a very modest growth rate.

Within the past ten thousand years, the idea of agriculture arose independently at several places. With agriculture came an explosion of technological development and social institutions and a gradual end to the gatherer/hunter existence. Agriculture produced food in abundance and the new social institution provided for its distribution. The population grew and concentrated in villages, towns, and cities. Between 8000 B.C. and 1 A.D. the population growth rate may have been 0.4 percent per year (Coale 1974).

Although humanity had largely given up the gatherer/hunter existence, there were still reasons to move. Curiosity, food shortages, overcrowding, war, religion, politics, and countless other factors motivated countless emigrations. Some of the post-agricultural settlement ventures have been conducted by political institutions — notably the European colonies in North America. These were true colonies rather than settlements in that the sponsoring institution maintained political control.

There have been two factors which have tended to assure that human settlement has largely remained a process driven by the decisions of individuals and/or small groups. First, even when ocean-going vessels were needed for transportation, they have never been so expensive that large institutions could control the means of emigration. Second, transportation and communication were slow enough that control of emigration or immigration could not be maintained for long. Human institutions seem not to be able to maintain control from a distance. Local institutions evolve to meet local needs. Colonies become independent communities after they achieve self-sufficiency.

We have spread across the face of Earth and planted permanent, self-sufficient communities on all but the most inhospitable lands. Of places on Earth, only the ocean floor remains to be settled. We have no other place left to go but into space.

SETTLEMENT OF THE SOLAR SYSTEM

In the near term, we will colonize rather than settle space. Proposed large orbital habitats (O'Neill 1974) and their support technologies could be established only at enormous expense. Only the largest of human institutions could conceivably pay the estimated 10^{11} cost of establishing the first orbital community of 10^4 colonists. Such an enormous enterprise may never be attempted. However, unless we abandon space ventures entirely, economic, scientific, and political justification should exist for establishing permanently manned facilities in space and, possibly, smaller installations on the surfaces of the rocky planets and moons.

The enormous cost of lifting mass from the surface of Earth and the other large bodies in the solar system will probably guarantee that the space-living population will grow slowly and that efforts will be made to make the orbital facilities self-sufficient in order to reduce the costs of resupply.

Gradually, the space-living population and the technology base in space will increase. Habitats built from asteroidal or lunar materials will house small communities. Clusters of communities may occur near raw materials (the Earth-Moon system, Mars, Jupiter, the Asteroidal zone), but humans will be scattered throughout the solar system, living in habitats connected by vast communications and data networks and separated by small energy differentials.

INTERSTELLAR SETTLEMENT

Once a sufficient base of population and technology is established in near-Earth orbit, the settlement of the solar system should proceed fairly steadily. Solar system distances are short enough that no journey would take more than a few years; the energy expenditures for travel would be modest. And, as in the pre-agricultural expansion of humanity across the face of the Earth, the crucial resources — sunlight, carbon, water, etc. — are accessible throughout the solar system, particularly among the asteroids and in the Jovian and Saturnian systems.

The next step outward, the step into interstellar space, introduces a new factor into the expansion/settlement process: extreme separation of potential settlement sites. Unless we assume that human communities can flourish in interstellar space (Dyson 1979), the settlement sites will be orbits close enough to stars that starlight can be used as the principle energy source and that a dependable supply of raw materials in the form of asteroids, comet-nuclei, moons, and gas giant atmospheres is near at hand. If we assume that settlements will be established in orbit about single, late-type stars and that the interstellar vessels move at about one-tenth light speed, the settlements will be separated, on the average, by 4 parsecs and journeys of 125 years (Jones 1978).

In no previous human expansion has the average journey been life long. Even though the interstellar vessels may house large communities of emigrants (perhaps 10^4 or 10^5 per vessel) and resemble the interplanetary habitats in which the emigrants and their ancestors had lived for centuries, millenia, or longer, and even though communications could be maintained with the home system and with the destination, the emigrants would be effectively isolated from the rest of humanity by distance and by communications time lags of up to 12 years! It is difficult to imagine the maintenance of interstellar colonies. Even if the journeys were made in suspended animation, a 125-year time gap could be a serious problem for colonial administrators. It seems evident that any human settlements established outside the solar system will, of necessity, be independent, self-sufficient communities.

There was an analogous human migration, recorded in part in the oral history of the Polynesian peoples (Buck 1938). In seemingly extraordinary feats of navigation, the widely separated islands of the Pacific were settled, perhaps in several episodes, by agricultural/fisher peoples of Asiatic origin. Although there is considerable debate on many details of the settlement of Polynesia, it appears that the Pacific was settled long after the rest of the world. Settlement of Easter Island, 1100 miles from the nearest land (Pitcairn), occurred only a few hundred years before the arrival of Europeans.

The Polynesians were superb navigators. They used the stars, winds and currents, and observations of migratory birds and clouds as they crossed hundreds of miles of open ocean between island groups. The initial discovery of an island may have involved some luck (good or ill) but, once found and reported, the islands were not lost.

The great Pacific distances required that the emigrants carry with them supplies for voyages of up to three or four weeks as well as everything they would need when they arrived at their destination. Along with themselves and their culture, they brought foodplants (notably coconut, bananas, and breadfruit) and domestic animals (dogs, fowl, and pigs). Contact was maintained over long periods with other parts of Polynesia, but cultural and language divergence was certainly evident by the time the Europeans arrived (Buck 1938). The Polynesian peoples and culture spread throughout the Pacific because the necessary technology was available. In some sense it was the technology that spread. Similarly, in the interstellar case, cultural and even genetic drift can be expected. If human descendants fill the Galaxy, they may well be members of diverse species and cultures but their biologic and technologic heritage will have originated in the solar system.

MATHEMATICAL MODELS

The dispersal of a tracer gas in a background gas is described by the diffusion equation. Each molecule of tracer moves an average distance, called the mean-free path, between collisions with the background gas. The collisions are isotropic; there is no preferred scattering angle. A requisite for application of the diffusion equation is that the mean-free-path is short compared with any gradient length scales; the density of tracer particles cannot change by large factors over a mean-free path.

There are clearly physical situations to which the diffusion equation does not apply. An obvious example is a chemical explosion in which the explosive products move outward at high speed, forming a shock wave. At the shock front the gradient length scale is comparable with the mean-free path, and the approximations which lead to the diffusion equation break down.

The dispersal of plant and animal species has been successfully treated with the diffusion equation (for example, Newman 1980). The modifications necessary to extend the diffusion model to living creatures are the addition of a source term (the population increases) and interpretation of the mean free path as the average distance between an individual's place of birth and the birthplace of offspring. As long as the gradient length scales remain relatively large, the modified diffusion equation is an entirely adequate description of the dispersal of life.

Newman and Sagan (1978) have attempted to apply the diffusion equation to interstellar migrations. Their equation may be written as

$$\frac{\partial P}{\partial t} = \alpha P \left(1 - P/P_s\right) + \gamma \Delta^2 \frac{\partial}{\partial x}\left(\frac{P}{P_s}\frac{\partial P}{\partial x}\right) \tag{1}$$

where

P = the population of a settlement,
P_s = the carrying capacity of a settlement,
t = time,
x = spatial coordinate,
α = local population growth rate,
γ = emigration rate, and
Δ = mean separation of settlements.

The solution to equation (1) is

$$P/P_s = 1 - \exp\left(\frac{x - \nu t}{L}\right) \tag{2}$$

where

$$L = \Delta\sqrt{\frac{2\gamma}{\alpha}} = \text{gradient length scale}$$

and

$$\nu = \sqrt{\frac{\alpha\gamma}{2}} = \text{wave speed.}$$

For problems of interest the local growth rate (α) greatly exceeds the emigration rate (γ) so that $L \ll \Delta$. The clear implication is that the modified diffusion equation cannot apply to interstellar migrations.

In some ways interstellar migrations may resemble explosions. With relatively slow emigration rates the populations at neighboring settlements can differ by large factors. Spatial population gradients can be steep at the frontier and important to calculations of an accurate solution.

A more appropriate solution method is a discrete, Monte Carlo simulation (Jones 1976, 1978, 1980). Briefly, we scatter settlement sites in a test volume. We assume that local population growth is given by

$$\left[\frac{\partial P}{\partial t}\right]_L = \alpha P \left(1 - P/P_s\right), \tag{3}$$

and that emigration is

$$\left[\frac{\partial P}{\partial t}\right]_E = \begin{cases} -\gamma P \left(1 - P/P_s\right) & P > P_L \\ 0 & P < P_L \end{cases} \tag{4}$$

where the parameter P_L is a threshold population such that emigration occurs when $P > P_L$, while immigration from nearby settlements occurs only when $P < P_L$. Numerical experiments demonstrate that the precise treatment used to describe emigration and immigration is relatively unimportant to the solution. More details are given by Jones (1976, 1978, 1980).

CALCULATIONAL RESULTS

Monte Carlo solutions have been obtained under these assumptions:

1. the density of settlements is 0.0015 pc^{-3}. This corresponds to a mean separation of sites of 2.2 parsecs;
2. the ship speed (v_s) is 0.1 c = 0.03 pc yr^{-1};
3. the maximum voyage is 22 parsecs; and
4. emigrants are sent alternately to the two nearest open settlement sites ($P < P_L$).

The solutions are not particularly sensitive to the choice of a ship range unless the range is comparable to the separation of sites. If this is the case, settlement is likely not to occur; the home system and/or the first few settlements are isolated from other potential sites. The settlement wave dies off.

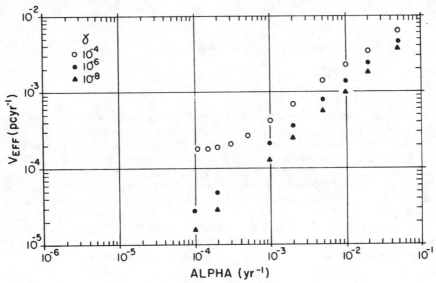

Fig. 9.1. The wave speed calculated for human settlement of the Galaxy is plotted as a function of the local population growth rate (alpha) for three values of the emigration rate (gamma). Newman (private communication) has shown that the wavespeed is well approximated by $v = 0.7 \, \alpha \, \Delta / \ln (2\alpha/\gamma)$, where Δ is the mean separation of settlement sites. For these calculations we assumed Δ = 2.2 parsecs. The history of human migrations and population growth suggests that reasonable values of the parameters are α = 10^{-3} and γ = 10^{-4}. The calculated wavespeed is 5 x 10^{-4} parsecs per year and the implied time needed for humanity to settle the Galaxy is 60 million years.

The solutions are presented in Fig. 9.1, as calculated wave speed as a function of the coefficients α and γ, both given in parts per year. The wave speed is given in parsecs per year. Newman (private communication) has shown that the Monte Carlo results are well approximated by

$$\nu = \overline{\delta r}/\left[(\overline{\delta x}/\nu_s) + (1/\alpha)\ \ln\ (2\alpha/\gamma)\right] \tag{5}$$

where

$\overline{\delta r}$ is the average radial distance traveled, and
$\delta x/\nu$ is the average travel time.

Usually we can assume $\overline{\delta r} = 0.7\Delta$ and neglect the travel time and use

$$\nu = 0.7\alpha\Delta\ /\ln\ (2\alpha/\gamma). \tag{6}$$

DISCUSSION

It is evident from Fig. 9.1 that as long as $\gamma << \alpha$, the wavespeed is dominated by the local growth rate (α). The evidence suggests that since the rise of agriculture the rate has exceeded 10^{-3} per year (Coale 1974). It seems likely that the very high rates experienced since the industrial/scientific revolution (1 to 2 percent per year) are a temporary departure from normality due to the dramatic decrease in the death rate in the past 200 years. We speculate that a more appropriate rate would be $\alpha = 10^{-3}\mathrm{yr}^{-1}$.

If $\alpha \geq 10^{-3}\mathrm{yr}^{-1}$, it is evident that the wave speed will exceed 10^{-4} pc yr unless the emigration rate is less than 10^{-8}. The choice $\alpha \geq 10^{-3}$ yr-1 gains some credence from the consideration that if one invokes small α to explain the absence of extraterrestrials from the solar system, then small growth rates must be invoked for all technological civilizations.

From data given by Potter (1965), the estimated emigration rate from Europe to North America during the 18th century was 3×10^{-4} yr-1. From data cited by Davis (1974), emigration in the period 1840 to 1930 often exceeded that value. The great Irish emigration of that period saw rates as large as 0.01 yr-1! About one hundred Irish arrived at American ports each day for years on end. It seems likely that $\gamma = 10^{-8}$ yr-1 as suggested by Newman and Sagan (1978) is a gross underestimate of human emigration rates.

Even if γ is as small as $10^{-8}\mathrm{yr}^{-1}$, the wave speed will exceed 10^{-4} parsecs per year. Since the Galaxy is a thin disk about 3×10^4 parsecs in diameter, we estimate that humanity could fill the Galaxy in no more than 300 million years. Using the more likely values of $\alpha = 10^{-3}$ yr-1 and $\gamma = 10^{-4}$ yr-1, we get a wave speed of 5×10^{-4} parsecs per year and a migration time scale of 60 million years. Because both of these times are short compared to the 4.6×10^9 year age of the solar system, the apparent absence of non-native civilizations from the solar system is significant. Only if $\alpha < 10^{-4}$ do the time scales become comparable.

CAVEATS

Are we alone? The settlement calculations suggest that we are. However, the calculations are based on extrapolations from our past and speculations about our future. They must be viewed with suspicion.

There are several assumptions we have made which could be drastically wrong. Foremost among these is that we assume interstellar travel is practical. Interstellar settlement will not occur on any significant scale if the voyages are very expensive. I suspect that if the labor costs exceed a few man-years per emigrant, the emigration rate will be very low, if not zero. The cost will not include the technological base. That will have to exist for its own sake. Interstellar transportation systems will have to be an outgrowth of ordinary interplanetary systems – interstellar 747's rather than interstellar Saturn V's. I suspect that the requirement that interstellar travel not be terribly expensive is not a serious prerequisite. Interstellar transportation systems may seem expensive from our perspective, but then, so would a 747 to the Wright brothers.

One further caveat needs to be mentioned. If humanity can live in interstellar space, using stars and planetary systems merely as way-stations (Dyson 1979), our estimates of the settlement times may be meaningless. The diffusion equation will apply, Δ will be very small, and most importantly, the absence of obvious signs of settlements in the solar system will be insignificant.

If I were forced to bet on the question, I would bet that we are alone. But we won't have a definitive answer until we have explored a large portion of the Galaxy and found no one home.

ACKNOWLEDGMENTS

This work has greatly benefited from several lively discussions and exchanges of letters and manuscripts with W.I. Newman. My colleagues at Los Alamos have also contributed thoughts and hours of patient listening: R.W. Whitaker, M.T. Sandford II, B.W. Smith, and H.G. Hughes. I gratefully acknowledge the continued interest and support of R.R. Brownlee.

The assistance of the staff of the National Library, Dublin, Ireland, is gratefully acknowledged. In search of an ancestor, my wife and I were directed to copies of 19th-century passenger lists. Our examination of the lists of people arriving at New York, Baltimore, and other American ports on a single day in August 1859 gave a sense of the magnitude of the Irish emigration. These lists are also available at the National Archives, Washington, DC.

I note that Peter H. Buck, whose book Vikings of the Pacific is an interesting introduction to the Polynesian migration, was Maori/Irish – hence, the product of two great migrations.

This work has been supported by the Los Alamos Scientific Laboratory which is operated by the University of California under contract to the U.S. Department of Energy.

REFERENCES

Buck, Peter H. (1938). Vikings of the Sunrise. J.B. Lippincott Co., New York, reissued as Vikings of the Pacific, University of Chicago Press, 1955.

Coale, Ansley J. (1974). The history of the human population. Sci. Amer., 231, 40-51.

Davis, Kingsley (1974). The migrations of human populations. Sci. Amer., 231, 92-105.

Dyson, Freeman J. (1979). Disturbing the Universe. Harper and Row, New York.

Hart, Michael H. (1975). An explanation for the absence of extra-terrestrials on Earth. Q.J. Royal Astron. Soc., 16, 128-135.

Jones, Eric M. (1976). Colonization of the galaxy. Icarus, 28, 421-422.

Jones, Eric M. (1978). Interstellar colonization. J. Brit. Interplanet. Soc., 31, 103-107.

Jones, Eric M. (1980). Discrete calculations of interstellar settlement. In preparation.

Newman, W.I. (1980). Some exact solutions to non-linear diffusion problems in population genetics and combustion, J. Theoretical Biology. In press.

Newman, W.I., and Sagan, C. (1979). Galactic civilizations: population dynamics and interstellar diffusion. Preprint, December 1978, revised August 1979.

O'Neill, G.K. (1974). The colonization of space. Physics Today, 27, 32-40.

Potter, J. (1965). The growth of population in America, 1700-1860. In Population in History (D.V. Glass and D.E.C. Eversley, Editors), 631-679. Edward Arnold Publishers, New York.

DISCUSSION

Michael Hart: I noticed that you used essentially a constant alpha in the computations. It would seem to me that at a time when a particular planet or colony had a low population they might want to have a rapid rate of growth, particularly if they were seeking colonists to encourage a new colony setting out. The limits on alpha then are nothing like 10^{-3} or 10^{-2} but can be very high figures indeed. If you wanted to artificially have a high rate of population growth just for a brief stretch of time, you could achieve rates of ten or twenty percent per year.

Jones: The dominant factor in the expansion rate is the population growth rate, so if you increase that you are going to increase the wave speed. The effort here was to make reasonable or conservative assumptions.

Hart: You used a rather short trip length, two or three parsecs?

Jones: Four parsecs.

Hart: An occasional long trip, even if only one trip in a thousand were as long as 50 or 100 parsecs, would distinctly increase the speed of colonization.

Jones: It turns out that it doesn't, unless you include very long trips. I've done a number of calculations with twenty parsec journeys; it increases the wavespread by 25 percent. You have to get very large.

Ronald Bracewell: What about the colonies that are left behind? Do they continue to send out colonies too in this model?

Jones: If there is an open site — that is, a site with population less than P_L — within ship range, yes, they will continue to send folks out. But if you limit the ship range to a few times the mean separation, then they very quickly find themselves back in the interior, isolated from the frontier.

Gerald Feinberg: A comment concerning immigration. As far as I know, in past examples the population of the home society never decreased through immigration. The population of Europe went up like a shot during the period that you were talking about, even though they were sending those people out. And one of the reasons for that is that they were sending people here and we were sending potato plants back there, which enabled more Europeans to live there.

Jill Tarter: What happens to the land-locked people who have no access to the frontier?

Jones: Since the wavespread is much less than the velocity of light, the folks on the frontier are talking with the folks back in the saturated interior. People are remarkably comfortable living in very strange places, like big cities for example. I don't understand that at all.

Harlan Smith: The answer to that question is that they may just decay in the interior. That reminds me of a story. On an eclipse expedition to Greenland, I encountered a Finnish geodesist from an adjoining camp. As we walked over the tundra he showed me a type of lichen on the rocks which was old and rotten at the center, but always expanding. And he said, "back home we call that lichen 'mother Russia'."

10

Colonies in the Asteroid Belt, or a Missing Term in the Drake Equation

Michael D. Papagiannis

ABSTRACT

The colonization of the Galaxy by advanced civilizations appears to be an inevitable cosmic event that can be accomplished in about 10 million years. The Drake equation suggests that the Galaxy contains now 10^5-10^6 advanced civilizations that originated and evolved independently of one another, which implies that close to 10^9 technological civilizations must have blossomed in the Galaxy during the past several billion years. It is highly unlikely, however, that the Galaxy could have hosted so many independent civilizations over billions of years and not yet have been fully colonized. Consequently, either the Galaxy has already been colonized, in which case space colonies must be in orbit around every well behaved star of the Galaxy including our own, or the colonization of the Galaxy has not yet taken place because actually there were only few, if any, other civilizations in the long history of the Galaxy to initiate the colonization process, in which case a correction is needed for the Drake equation. The asteroid belt, an excellent source of raw materials, seems to be the best place to search for space colonies, while the missing term of the Drake equation is probably related to the difficulty of maintaining a clement environment on a planet for several billion years so that life will have the necessary time to evolve from primitive micro-organisms to advanced civilizations.

THE DRAKE EQUATION AND THE COLONIZATION OF THE GALAXY

The Drake Equation (Drake, 1965; Shklovskii and Sagan, 1966) tries to estimate the number N of advanced civilizations currently present in the Galaxy from the expression,

$$N = R \cdot P \cdot L \tag{1}$$

where R is the rate at which stars are being formed in the Galaxy, P is the probability that a given star will have all the necessary conditions (planets at the right distances, etc.) for the origin and subsequent evolution of life to advanced civilizations, and L is the average lifetime of advanced civilizations. The values commonly assigned to these three factors (Papagiannis, 1978a; Goldsmith and Owen, 1980) are,

$$R \simeq 20 \text{ stars/year} \tag{2}$$

$$P \sim 10^{-2} \tag{3}$$

$$L \simeq 10^6 \text{ years} \tag{4}$$

so that the number N of advanced civilizations currently present in the Galaxy becomes,

$$N \sim 10^5 - 10^6. \tag{5}$$

Since the age of the Galaxy is much longer than the average lifetime of advanced civilizations, it follows that a much larger number N_T of advanced civilizations must have appeared throughout the history of the Galaxy. This number N_T is given by the expression (Freeman and Lampton, 1975; Papagiannis, 1978a)

$$N_T = R \cdot P \cdot T \tag{6}$$

where T is the period over which civilizations could have been appearing in the Galaxy. Allowing 2-3 billion years from the formation of the Galaxy for the production of heavy elements and 4-5 billion years from the formation of a solar system to the appearance of an advanced civilization, we still have,

$$T \simeq 5 \times 10^9 \text{ years} \tag{7}$$

and hence the total number N_T of advanced civilizations that must have appeared in the Galaxy during the past 5 billion years becomes,

$$N_T \simeq 10^9 . \tag{8}$$

The Drake equation, of course, excludes the possibility of stellar colonization and assumes that each of these civilizations arose independently, starting spontaneously as primitive life on a hospitable planet and slowly evolving into a technological civilization. In view, however, of our more recent ideas about galactic colonization, these results represent a contradiction in terms, because it is extremely unlikely that nearly a billion civilizations, each one with an average life of about a million years, could have existed in the Galaxy for several billion years without any of them initiating the colonization process, a process which in a mere 10 million years would have converted the Galaxy into a fully inhabited world.

Our present views on the colonization of the Galaxy are based to a great extent on the ideas of human colonies in space advanced in recent years by O'Neill (1977) and others. In the beginning these space colonies will probably be small scientific stations with a modest number of scientists and crew, but as time goes on they will become large habitats housing thousands of people on a permanent basis. In time they also will become totally independent of the Earth, getting their energy either from the Sun or from fusion, and obtaining all their raw materials from space and most likely from the asteroid belt. As these colonies proliferate, many will cut completely their ties with the Earth and will set themselves up in independent, more advantageous orbits around the Sun that will bring them closer to their supplies of energy and sources of raw materials.

Such colonies would be capable of undertaking long trips to other stars at moderate speeds of $V \sim 0.01-0.05$ c, which are entirely feasible from an energy point of view with the use of nuclear fusion (Hart, 1975; Papagiannis, 1978a,b). These interstellar trips, of course, will take several hundred years and hence many generations; but this will be of no great concern to their inhabitants, since their lives will continue to follow the same patterns as when they were orbiting aimlessly around the Sun. Our own planet, after all, is a colossal spaceship that orbits the Galaxy in about 250 million years, and nobody seems to mind this grand trip.

Space colonies will also make the colonization of the Galaxy much easier and much more complete, because their inhabitants will prefer to continue to live in space habitats, as they did for many generations during their long interstellar trips, rather than search for planets with an Earth-like atmosphere to settle. All they would need, therefore, in a new solar system would be raw materials to build and sustain more space colonies. Raw materials, however, are likely to be available in practically every solar system of the Galaxy and hence the colonization is likely to engulf every well behaved star of the Galaxy, including of course our Sun. Conversely, the colonization process would have been greatly hampered if the colonizers were able to settle only on planets with an Earth-like atmosphere, because planets with an atmosphere so similar to ours that people could walk out of their spaceships and breathe freely in the open air are likely to be extremely rare, popular movies and television programs not withstanding.

The reason is simply that in order for a planet to have an atmosphere very similar to ours (and the tolerance as we all know from climbing a high mountain is rather small), it must have a mass and spinning rate similar to that of the Earth and must be at nearly the same distance from a star very similar to our Sun in mass, chemical composition and age (Ulrich, 1975). In addition, life on it must have followed a similar evolutionary path, including photosynthesis, and it must now be at a similar stage of its evolutionary process. All these stringent requirements are bound to make such coincidences in the Galaxy quite infrequent. The colonization of the Galaxy, therefore, would have been very difficult for colonizers searching for Earth-like planets and also very incomplete, because the few available sites for colonization would probably be hundreds of light years away from each other. All these problems, however, are nicely solved with the use of space colonies.

It must be mentioned here that several prominent scientists, including Drake (1980) and Newman and Sagan (1980), have raised different objections to interstellar travel and to the feasibility of galactic colonization. An analysis (Papagiannis, 1980a), however, of all these technical, economic and social problems shows that, though any one of these difficulties might prevent some galactic civilizations at some stage of their development from initiating interstellar travel, it is virtually impossible to find a universal reason that would prevent each of a billion galactic civilizations from doing so over a quadrillion years of combined existence. It seems unavoidable, therefore, especially in view of the natural tendency of life to want to expand to occupy all available space (Papagiannis, 1980a), that in the presence of a large number of stellar civilizations interstellar travel will be initiated by at least some of these civilizations and that the colonization wave will sweep through the entire Galaxy in about 10 million years (Hart, 1975; Papagiannis 1978a,b; Drake, 1980).

SPACE COLONIES IN THE ASTEROID BELT

If the Galaxy has already been colonized, the number of advanced civilizations in the Galaxy must now be in the 10^{10}-10^{11} range, with practically every reasonable star populated with space colonies. The likelihood that the colonization is now in progress is very low (~0.2 percent), because the colonization time (~10^7 years) is much shorter than the period (T~ 5×10^9 years) over which civilizations must have been capable of initiating the colonization process. There is also no way that the colonization wave would have missed, or by-passed our own solar system, which meets all the essential requirements for the establishment of space colonies (longevity, stability, a bright but not too bright sun, and a multitude of planets, moons, comets and asteroids). The reason is that every colonized star becomes in turn a base for future missions and thus in a very short interval every reasonable star will be colonized by its neighbors.

The question then is: If our solar system has already been colonized, where are all these colonies located? As discussed above, it is unlikely that the space travellers would find planets on which they would be willing to settle, and therefore what we should be looking for are space colonies. But where are these space colonies likely to be found? If solar energy was the most important commodity for the space colonies, one would expect to find them in orbits closer to the Sun. These people, however, must have already solved their energy problems, most probably through the use of nuclear fusion, because otherwise they would have not been able to undertake interstellar trips of several centuries away from any star. Consequently, their most important commodities must be raw materials including probably deuterium, tritium and possibly He^3 (Martin and Bond, 1980) for their nuclear burners. The asteroid belt, therefore, appears to be the most logical place for these space colonies (Papagiannis, 1978b), because it can provide them with all the necessary materials, from metals to organic compounds, at a very low energy expense for their retrieval. In addition nuclear fuels for fusion can be obtained from Jupiter, while at a distance of 2-3 A.U. from the Sun solar energy could still be a viable supplementary source of energy if needed.

Space colonies in the asteroid belt could have easily escaped detection up to now, lost among the myriads of natural asteroids of similar sizes. A more thorough search, however, especially designed for this purpose, might be able to detect certain physical anomalies in some asteroids, such as temperatures considerably higher than those expected in terms of their distances from the Sun, that would indicate an artificial origin. It would be a great mistake to search in thousands of far-away stars for signs of intelligent life without looking also into our own back yard where detection and contact would be much easier.

A MISSING TERM IN THE DRAKE EQUATION

If a careful search in our solar system and in the nearby stars should reveal no trace of space colonies, we would have to conclude that the Galaxy has not yet been colonized. This can be explained only if the number N_T of advanced civilizations that have appeared throughout the history of the Galaxy is quite small, i.e., if $N_T < 10^3$. The terms R and T, however, in (6), which gives N_T, cannot be changed by any large factor. In order, therefore, to change N_T from about 10^9 (8) to less than 10^3, the only alternative is to change the probability P from about 10^{-2} (3) to less than 10^{-8}. The probability P used in the Drake equation is actually defined as a product of 6 terms (Papagiannis, 1978a),

$$P = f_g \cdot f_p \cdot n_e \cdot f_l \cdot f_i \cdot f_a \qquad (9)$$

where f_g is the fraction of stars that are quite similar to our sun, and is typically assigned values (Hart, 1979; Pollard, 1979) of the order of

$$f_g \approx 10^{-2} \qquad (10)$$

while all the other terms are usually given values close to unity (Goldsmith and Owen, 1980),

$$f_p \simeq n_e \simeq f_l \simeq f_i \simeq f_a \simeq 1 \qquad (11)$$

though several other variations have also appeared in the literature (Sturrock, 1980).

The fraction f_p of stars that have planets is not known, and hence neither is the number of planets n_e found inside the ecosphere, or habitable zone, around these stars. We are essentially extrapolating from our own example when we assign values close to unity to these two terms. We know even less about the terms that give the fraction of planets where life originates f_l, evolves into high intelligence f_i, and finally reaches a level of advanced technology f_a. The example of the Earth indicates that the spontaneous origin of life in a hospitable planet might not be very difficult since in our case it originated at least 3.5 billion years ago (Lowe, 1980), i.e., only 0.1-0.7 billion years after the formation of the oceans. The evolution, on the other hand, of life to high intelligence and to advanced technology seem inevitable, since both endow the species that possesses them with distinct advantages. The problem, however, is that it takes several billion years of planetary and biological evolution to reach these final stages. In the case of the earth, e.g., the evolution to high intelligence took 5-40 times longer than it took for life to originate. It appears, therefore, that the painfully slow evolution to high intelligence represents the bottleneck of the entire process.

The Drake Equation neglects the factor of time, assuming presumably that there is always plenty of time available. This is not so, however, for evolutionary periods of the order of 4 billion years which represent roughly 30 percent of the age of the Galaxy and 85 percent of the life-time of our Sun. Hart (1978, 1979) has shown that the increase in the luminosity of the Sun and different evolutionary changes in the composition of the atmosphere of a planet can easily lead either to a runaway greenhouse effect, as in the case of Venus, or to runaway glaciation, as possibly in the case of Mars. It seems, therefore, that many planets which originally, or at a given stage of their evolution, satisfied the requirements for the origin of life, will not be able to maintain these favorable conditions (essentially, liquid water on their surface) for several billion years so that life might be able to evolve to an advanced civilization.

In addition to the increasing luminosity of the central star (Ulrich, 1975), there are also many factors related to the planet which can lead to runaway glaciation or to a runaway greenhouse effect, thus eliminating liquid water from the surface of the planet. These factors (Papagiannis, 1981) include the mass of the planet, the rate of rotation, the tilt, precession and nutation of the axis of rotation, the eccentricity of the orbit, different secular variations of the orbital parameters, the presence of a magnetic field, the presence of a large moon, the presence of a ring around the planet, continental drift, plate tectonics, etc., all of which can disturb the delicate balance between the amount of incoming radiation, albedo and greenhouse effect which allows the planet to maintain a moderate average temperature. If this balance is disturbed, the temperature will move toward either extreme; this will eliminate liquid water from the surface of that planet and will bring the evolution of life on that planet to an end. Of course, one can always hypothesize about other forms of life that thrive without water, but objective arguments (Goldsmith and Owen, 1980) favor a carbon and water based life as by far the most likely possibility in the Universe.

From the above it follows that only a very small fraction of planets where life originates might be able to offer the long term (billions of years) stability required for the slow evolution of life to high intelligence. This time effect can be incorporated into the Drake equation by including an additional term f_e which represents the fraction of planets where life will be given the necessary time to complete the inevitable, but extremely slow evolutionary process to high intelligence. The revised probability p' with this new term f_e included, will then be

$$p' = Pf_e = f_g \cdot f_p \cdot n_e \cdot f_l \cdot f_e \cdot f_i \cdot f_a \tag{12}$$

This evolutionary term, however, might prove to be extremely small,

$$f_e < 10^{-6} \tag{13}$$

which would make the revised probability p' very small too,

$$p' < 10^{-8}. \tag{14}$$

When this new probability p' is used in (6), it produces a small value for N_T,

$$N_T < 10^3 \tag{15}$$

which can then explain the lack of colonization in the Galaxy.

CONCLUSIONS

By accepting the principle that the presence of a large number of advanced civilizations over billions of years makes inevitable the colonization of the Galaxy, one is led to two, diametrically opposed, alternatives. The one is that the Galaxy has already been colonized and there must be space colonies in orbit around every well-behaved star of the Galaxy including our Sun. In this case, the most likely place to look for these space colonies in our own solar system seems to be the asteroid belt (Papagiannis, 1978b).

The other alternative is that the Galaxy has not yet been colonized because of the lack of a substantial number of advanced civilizations in the Galaxy during the past several billion years. The reason could very well be that very few planets manage to maintain liquid water on their surface for billions of years (Papagiannis, 1981) to accommodate the extremely long evolutionary process from a primitive micro-organism to an advanced civilization. In this case, an evolutionary term $f_e < 10^{-6}$ must be added to the Drake equation that would reduce by at least 6 orders of magnitude the number of advanced civilizations that have appeared in the Galaxy and would thus explain the lack of galactic colonization.

The rapid development of our scientific and technological capabilities (Papagiannis, 1980b) will probably make it possible within the coming decades to decide which one of these two opposing alternatives represents the real world.

REFERENCES

Drake, F.D. (1965). In Current Aspects of Exobiology, (editors: G. Mamikunian, and M.H. Briggs). Pergamon Press, New York.

Drake, F.D. (1980). N is neither very small nor very large. In Strategies for the Search for Life in the Universe, (editor: M.D. Papagiannis). D. Reidel, Dordrecht, Holland.

Freeman, J. and Lampton, M. (1975). Interstellar archaeology and the prevalence of intelligence. Icarus 25, 368.

Goldsmith, D. and Owen, T. (1980). The Search for Life in the Universe. Benjamin/Cummings, Menlo Park, California.

Hart, M.H. (1975). An explanation for the absence of extraterrestrials on Earth. Quarterly J. Royal Astronomical Soc. 16, 128.

Hart, M.H. (1978). The evolution of the atmosphere of the Earth. Icarus 33, 23.

Hart, M.H. (1979). Habitable zones about main sequence stars. Icarus 37, 351.

Lowe, D.R. (1980). Stromatolites 3400 million years old from the archaean of Western Australia. Nature 284, 441.

Martin, A.R. and Bond, A. (1980). Starships and their detectability. In Strategies for the Search for Life in the Universe, (editor: M.D. Papagiannis). D. Reidel, Dordrecht, Holland.

Newman, W.I. and Sagan, C. (1980). Galactic civilizations: Population dynamics and interstellar diffusion. Icarus (in press).

O'Neill, G.K. (1977). The High Frontier - Human Colonies in Space. William Morrow, New York, N.Y.

Papagiannis, M.D. (1978a). Could we be the only advanced technological civilization in our galaxy? In Origin of Life, (editor: H. Noda). Center Acad. Publ., Tokyo, Japan.

Papagiannis, M.D. (1978b). Are we all alone, or could they be in the asteroid belt? Quarterly J. Royal Astronomical Soc. 19, 277.

Papagiannis, M.D. (1980a). The number of galactic civilizations must be either very large or very small. In Strategies for the Search for Life in the Universe, (editor: M.D. Papagiannis). D. Reidel, Dordrecht, Holland.

Papagiannis, M.D. (1980b). Conclusions and recommendations from the Joint Session. In Strategies for the Search for Life in the Universe. (editor: M.D. Papagiannis). D. Reidel, Dordrecht, Holland.

Papagiannis, M.D. (1981). Liquid water on a planet over cosmic periods. In Proceedings of the 6th International Conference on the Origins of Life, (editor: M. Paecht-Horovitz). D. Reidel, Dordrecht, Holland.

Pollard, W.G. (1979). The prevalence of earth-like planets. Amer. Scientist 67, 654.

Shklovskii, J.S. and Sagan, C. (1966). Intelligent Life in the Universe, Holden Day, San Francisco, California.

Sturrock, P.A. (1980). Uncertainties in estimates of the number of extraterrestrial civilizations. In Strategies for the Search for Life in the Universe, (editor: M.D. Papagiannis). D. Reidel, Dordrecht, Holland.

Ulrich, R.K. (1975). Solar neutrinos and variations in the solar luminosity. Science 190, 619.

DISCUSSION

Gerald Feinberg: While it seems to be the case that it took four billion years for life to go from its origin on Earth to intelligence, we don't understand whether that number is a necessary one. We don't understand why life sort of dawdled along with one cell for three billion years and then suddenly took off to a multicell. It's conceivable that that number is also an accident, that in other circumstances you could go from the origin of life to intelligence in a much shorter period of time.

Papagiannis: I don't deny that it might be done either slower or faster. However, there are steps that are dependent on natural changes that you can't do without. For example, to go from unicellular to multicellular organisms you have to wait for the atmosphere to change from non-oxidizing to oxidizing. It seems that you need a different type of chemistry to have multicellular organisms; probably you need to depend on respiration rather than fermentation.

Feinberg: There seems to be some billions of years between the atmosphere becoming oxidizing and the multicelled organisms appearing. It's also not clear that it has to take two billion years for the atmosphere to become oxidizing if it starts off in the reducing state. That depends on the density of blue-green algae, which is very hard to predict from first principles.

Jill Tarter: I think it's wrong to treat life as sort of responding passively to external stimuli. The most profound effect on the atmosphere that you showed is the existence of life, and it had more effect than the nutation of the rotation axis or any other perturbation that you mentioned.

Papagiannis: Well, up to a certain point life can compensate for change. However, there is nothing much that an organism can do if you take the water out of it – at least the kind of life we know.

Unidentified speaker: Lovelock and Margulis' Gaia hypothesis argues that the biosphere has a whole feedback relationship with the gases in the atmosphere and so on. The temperature control would be a product of this interaction, and life wouldn't just be sitting there passively once it has its origin in the early Earth. They don't make a quantitative case for it, granted, but I think the suggestion of the remarkable event or process that you are talking about supports the idea that there is this self-regulation.

Papagiannis: I am willing to accept that there is this interaction. But you are also probably willing to accept the idea that Mars, once upon a time, had running water, as some of these pictures seem to indicate (although it's not proved yet), and now it doesn't. If there was life on Mars when there was liquid water, there wouldn't have been much that life could have done to prevent the water from completely freezing or evaporating.

Cliff Singer: The Gaia hypothesis had to be true here on Earth in the past or we wouldn't be here to observe it. Gaia is only a description of what happened on Earth. But there is no predictive ability for other systems. We have evolved in this band of fluctuating external pressures and fluctuating internal responses.

James Oberg: It's like you are a drunk wandering back and forth across a superhighway and cars are going back and forth. You make it across finally and you say, "golly, that's a good way to cross the road, I made it safely." And yet if you hadn't made it safely you never would have been able to say that. You don't know how many drunken bodies are lying back on the road. If thousands of planets started off Earthlike four billion years ago, how many would have made it? That's why I don't find the Gaia hypothesis too convincing.

11

Primordial Organic Chemistry*
Cyril Ponnamperuma

"If we could conceive, in some warm little pond, with all sorts of ammonium and phosphoric salts − light, heat, electricity etc. present, that a proteine compound was chemically formed ready to undergo still more complex changes. . . ." (<u>Charles Darwin to his friend Hooker, 1871.</u>)(1)

Here in a nutshell is the entire concept of chemical evolution. What the experimentalist does in the laboratory is to try to recreate Darwin's warm little pond and to see whether those reactions that preceded the emergence of life can be retraced in the laboratory. Such ideas lay fallow for a long period of time until the Russian biochemist Alexander Oparin, in a dissertation published in Russia in 1924, contended that there was no fundamental difference between a living organism and lifeless matter and that the complex combinations, manifestations and properties so characteristic of life must have arisen in the process of the evolution of matter.(2) In 1928, Haldane had similar ideas. He described the formation of a primordial broth by the action of ultra-violet light on the Earth's primitive atmosphere.(3) The Oparin-Haldane hypothesis is the basis of the scientific study of the origin of life.

ORGANIC SYNTHESIS

The composition of the sun gives us a reasonable blueprint for the raw material available for primordial organic synthesis. It is indeed the average composition of the solar system. Hydrogen, oxygen, nitrogen and carbon are most abundant and are the very elements that constitute

*Reprinted from <u>Chemistry in Britain,</u> 15, 560-68, 1979, with permission of the author.

over 99 percent of the biosphere. The primitive atmosphere of the Earth is considered to be a secondary feature which resulted from the outgassing of the Earth and was nonoxygenic (anoxic) in nature. Some components of the pristine atmosphere may have also come from later accretion(4) or even cometary sources.(5) During the prelife stage of the Earth the atmosphere must have been largely made up of methane, nitrogen and water and small amounts of phosphine and hydrogen sulphide. The energy sources available for primordial synthesis are ultraviolet light from the sun − especially in the short wavelength regions, which could have dissociated the components of the primitive atmosphere − electrical discharges, radioactivity in the crust of the Earth, heat from volcanoes and energy from shock waves.(6)

Over the years, several experiments have been performed with the objective of studying the effect of these forms of energy on the primitive atmosphere.(7) An early experiment, often overlooked, is that performed by Calvin, who in 1951 irradiated a mixture of carbon dioxide and water in the cyclotron at Berkeley, California, and was able to produce a number of organic compounds.(8) This experiment was followed by that of Urey and Miller in 1953.(9) The exposure of methane, ammonia and water to an electrical discharge gave rise to four amino acids: glycine, alanine, aspartic and glutamic acid. Since that time innumerable other investigators have attempted to retrace the path from a primitive atmosphere to nucleic acids and proteins.

To the chemist, prebiotic synthesis appears as a two-part problem: (i) to make the small molecules necessary for life; (ii) to combine the small molecules under similar conditions into the polymers, the poly-peptides and the oligonucleotides, which are the precursors of nucleic acids and proteins.

A typical laboratory apparatus is illustrated in Fig. 11.1. The upper flask represents the atmosphere, the lower flask the ocean. The side arm is hot, the condenser is cold. Circulation is thus established, portraying the interaction between the atmosphere and the hydrosphere. In this particular apparatus an electrical discharge is used. This energy source could be replaced by ionising radiation, by heat, by shock waves or by uv light. At the end of a 24 h experiment, 95 percent of the starting methane has been converted into organic compounds.   By careful analysis of the dark brown material which has been deposited on the flask and dissolved in the water, a number of molecules of biological significance can be isolated.(10)

In recent experiments a very powerful uv light source consisting of a 15 atmosphere (1.52 MPa) argon plasma with a total integrated dose of 10^{19} photons $cm^{-2} s^{-1}$ at a wavelength of 100-200 nm has been used. This flux is 10^4 times greater than that of the sun in that wavelength, and thus, in a short period of time, the energy of the sun over an extended duration could be simulated.(11)

In a typical experiment using ionising radiation, a long horizontal tube (Fig. 11.2) replaced the round flask used in the electrical discharge experiment.   In one of the first experiments performed with this apparatus adenine was synthesised. Not only was adenine formed, but

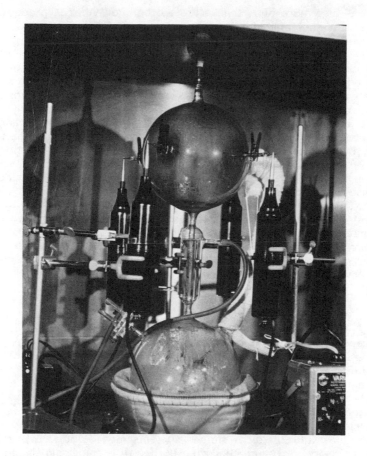

Fig. 11.1. Laboratory representation of the atmosphere and the ocean.

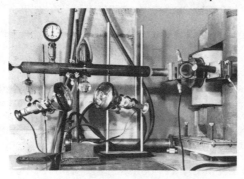

Fig. 11.2. Irradiation of a primitive atmosphere with an electron beam.

there was also more of it than all the other non-volatile products put together. At the time of this synthesis we were surprised, but later it was observed that hydrogen cyanide was synthesised in the process: adenine is the pentamer of hydrogen cyanide. The stepwise process from hydrogen cyanide to adenine, photo-chemically(12) or by the base catalysed reaction, has been established.(13)

With different types of energy acting on a primitive atmosphere, whether of an intensely reducing or non-oxidising nature, both hydrogen cyanide and formaldehyde appear to be formed. The importance of these compounds has been highlighted by their discovery in the inter-stellar medium. Hydrogen cyanide leads to the bases(14) while the pathway from formaldehyde to the sugars is analogous to the Butlerow reaction of old, where the base catalysed reaction of formaldehyde gave rise to a whole series of carbohydrates.(15)

In examining the possible pathway for the synthesis of amino acids it appears that the Strecker synthesis may have been effective. The combination of aldehyde and a nitrile gives rise to an alpha-aminonitrile which on hydrolysis gives rise to an amino acid. A simple scheme of organic chemistry can explain the origin of most of the molecules necessary for the proteins and the nucleic acids.

POLYMERISATION

The transition from the monomers to the polymers is mediated by a dehydration condensation. Such a dehydration could have taken place on the primitive Earth when the mixture of organic compounds, brought to the ocean shoreline and absorbed on a clay surface, was acted upon by solar heat. Bernal in his classic book, The physical basis of life, suggested that organic compounds from the ocean could have been polymerised through the catalytic effect of clay surfaces (see discussion below and in Cairns-Smith(43)). This idea of polymerisation on the ocean shoreline, or on the dried up bed of a primordial lagoon, has been successfully tested in the laboratory. However, it would be more plausible if such reactions could readily take place under aqueous conditions, since the Earth is a very wet planet (see Andrew(44)). Such a situation might give rise to a much more general explanation of the origin of polymers.

In attempting to simulate the dry ocean bed a nucleoside and a phosphate were mixed together and heated to about $125^{\circ}C$, whereby a large number of compounds were synthesised. Among them were found a dinucleoside monophosphate, mononucleotides, dinucleotides and tri-nucleotides.(16) More recent work using calcium phosphate as a phos-phorylating agent enabled us to synthesise polymers of up to 10 units.

In examining the polymerisation reaction under prebiotic conditions, it is clear that a number of condensing agents could have given rise to these results.(17) Prebiotic condensing agents that have been used are illustrated in Figure 11.3.(18)

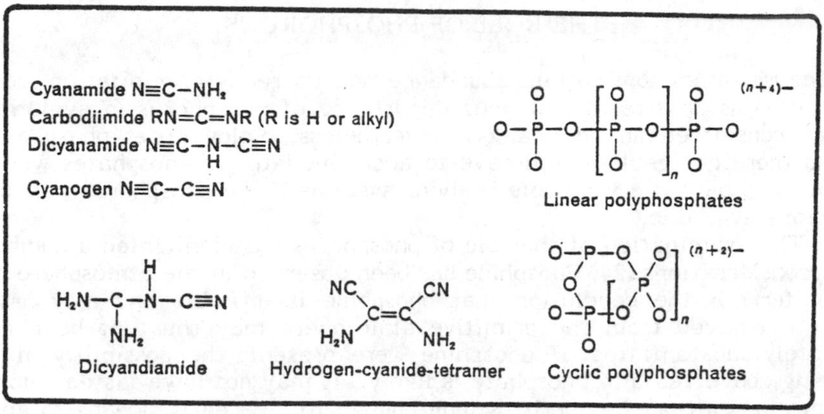

Fig. 11.3. Prebiotic condensing agents used in recent studies.(18)

Cyanamide is the simplest of the reagents that have been used. It is formed by the irradiation of a primitive atmosphere.(19) In one of the first experiments done under aqueous conditions glycine and leucine were irradiated with uv light in the presence of cyanamide. Among the products diglycine, triglycine, lucylglycine and glycil-leucine were identified.(20)

In other experiments carbodiimides have been used. The solvents here are non-aqueous in nature and may not have a direct bearing on primordial. chemistry.(21) Calvin and Steinman have examined the possible role of dicyanamide in prebiotic synthesis and shown that polypeptides could be formed at a low pH by such a reaction.(22)

In one of our earlier investigations of the action of an electrical discharge on a primitive atmosphere, no amino acids were detected by paper chromatography: when the resulting solution was hydrolysed, 10 amino acids were identified. It could be argued that the amino acids were a result of the hydrolysis of nitriles. When enzymatic hydrolysis was used, a number of amino acids were released, suggesting the presence of the peptide bond. The peptide bond appeared to be already formed after a 24 h experiment.(23) An interesting conclusion from this result is that a long period of time was not necessary for chemical evolution. If the right molecules were present, if organisation took place, then the appearance of life could perhaps be 'instant' in geo-chemical terms.

Since 15 percent of the primordial soup is made up of hydrogen cyanide it seems likely that the hydrogen cyanide, especially its tetramer, diaminomaleonitrile, may have been responsible for polymerisation in the prebiotic soup. Several experiments were designed using the hydrogen cyanide tetramer, clearly establishing its role in prebiotic chemistry, not only as a pathway for adenine but also as a condensing agent.

THE ROLE OF PHOSPHOROUS

Because of its low cosmic abundance and its ready precipitation from solution as apatite, the obvious importance of phosphorus in evolution was considered an engima.(24) Nevertheless, biology uses phosphorus and therefore geology will have to accommodate it. Phosphates would not have played a vital role in living systems if the phosphorus was not readily available.

The examination of the role of phosphorus has highlighted a number of considerations.(24) Phosphine has been observed in the atmosphere of Jupiter(25); the conclusion that phosphine is unstable and may have been removed from the primitive atmosphere may thus not be completely substantiated. If phosphine were present, the possibility of it being converted into phosphate is likely. It may not have lasted a long time in solution, but perhaps long enough for the early oceans to give rise to linear phosphates, some of the most effective condensing agents.

Another persuasive argument is that, if phosphorus were present in one form or another in the crust of the Earth, then during the course of heating generated by planetary accretion, polyphosphates would have been produced. If the oceans were formed by the outgassing of the crust these polyphosphates would have leached into the oceans and would have been productive as condensing agents. A further argument that comes to mind from the discussion of planetary formation is a contribution from late accretion. Even after the oceans were formed there was much scavenging in the solar system. Some of the falling debris from the primordial dust cloud, the mineral schribezite for example, would have provided some phosphides. The water reacting with phosphides would give rise to phosphine, and thus to phosphate and to polyphosphates.

In related experiments on the synthesis of organic compounds, when a suspension of clay in the form of sodium montmorillonite was used with a methane-nitrogen-water mixture and exposed to an electrical discharge, a number of interesting results were observed.(26) The clay plays several important roles in the synthesis, in catalysing the reaction of micromolecules, in absorbing them, protecting them from destruction, concentrating them and selecting some from the others, for example separating the protein from the non-protein amino acids. A specific result in this synthesis was the enhancement of higher molecular weight amino acids. Glycine is generally the most abundant amino acid produced in electric discharge experiments. However, in the presence of clay there appears to be more alanine than glycine. The further possible role of clay is an intriguing one, and is discussed by Cairns-Smith(43). Could clay have been responsible for the ordering of nucleotides and thus give rise to the process of gene formation?

Recent studies related to the radiation of acetic acid have shown that a number of carboxylic acids were formed.(27) These are important in the biochemical cycle such as the Krebs cycle. If we establish the transition between the individual carboxylic acids, we have the beginnings of a new concept in chemical evolution. The discovery, in a single experiment, of a number of carboxylic acids such as those that function in the Krebs cycle leads us to believe that not only individual molecules but also biochemical pathways antedated life.

The next stage in the study of chemical evolution revolves around interaction between nucleic acids and proteins. Under the most primitive conditions there must have arisen some possible relationship between polynucleotides generated under abiogenic conditions and polypeptides similarly formed. Today's genetic code is a complex apparatus that requires several steps from transcription to translation.   But perhaps, during the very early conditions, there may have been some relationship between the amino acids and mononucleotides, dinucleotides or trinucleotides. It is possible that the code may have been the result of the possible relationship between a particular set of nucleotides and an individual amino acid? In order to explore such a possibility several experiments have been performed.[28] A series of ion exchange columns, prepared with amino acids linked to them by covalent bonds, have been used as templates by which a selective absorption of nucleotides could be made. The specificity of the nucleotides for a particular amino acid can be noted.

In other experiments the relationships of nmr spectra of mono- and dinucleotides in the presence of a particular amino acid have been studied. The modification of the proton shift by the amino acids may be an indication of a potential relationship.[29]

Yet another approach is to look at the possible templates or fragments of polynucleotides which can specifically direct synthesis of activated amino acids. Such studies are the beginning of a programme to understand the next stage in chemical evolution, the transition from the organic to the biochemical.

In related theoretical studies an evolutionary picture of the early protein synthesis system appears to be emerging.[30] A model of a catalytic system has been studied in this connection. The system included a template, nucleotides and two activated amino acid polymerases. Variables defining the system were: catalytic activity of the polymerases, the number of the amino acid residues at the activity site, the number of amino acid residues at the selectivity site, the number of the polymerases, the accuracy of polymerisation and the activity of the polymerases, the number of evolutionary improvements, and the probability of occurrences of beneficial mutations. Computer calculations have indicated that even a small 'insignificant' activity and specificity of the polymerases could eventually lead the system to the most accurate protein synthesis.

The model study would encourage further quantitative investigations on catalytic activities of synthetic peptides and the interaction between nucleotides and amino acids and the construction of catalytic systems from a chemical evolutionary point of view.

EVIDENCE

The experiments that have been described so far would appear to be more credible if we had some way of finding out whether some of the primordial soup, or some of the molecules from the early prebiotic era, could be found hidden away in the crust of the Earth or on a planetary surface since the early stages of the evolution of the solar system. The studies by micropalaeontologists have taken us back some 3500 million years (My), where the presence of microfossils in sediments from South Africa and Western Australia have indicated the presence of life(29, 30, 45). Since the Earth is ca 4600 My old, we are particularly interested in sediments of the first 10^9 years of the Earth's existence, where organic matter of primordial nature may be preserved for us. Our search for these ancient sediments, especially those which may be described as 'molecular fossils' has taken us back 3800 My to the Isua supracrustal formation of western Greenland(31). Here there are sedimentary rocks which are so far the oldest known on the Earth. They contain graphitic material and, although they have been metamorphosed by various earth-movements, organic molecules have been found in them. The isotopic ratio is strongly suggestive of a flourishing biota. While these observations need further confirmation, we are in the position of asserting that life on Earth is about 3800 My old or that life is at least as old as the oldest known sediments(32).

Searches for prebiological matter have also been made in the lunar samples. Each of the samples from Apollo 11 to Apollo 17 has been analysed in our laboratory. An extensive search for traces of organic molecules which may be suggestive of chemical evolution has been made. Most of the work done points out that there are about 200 ppm of carbon in the samples, and that there is no evidence for any molecules of organic significance.(33)

Turning from lunar samples to meteorites, we have had the opportunity of examining several carbonaceous chondrites. This category of meteorites has carbonaceous material ranging from 0.5 percent to about 5.5 percent.

The Murchison meteorite fell in Australia on 28 September 1969. A fragment of that sample was analysed after its fall, and for the first time we were able to establish the presence of equal amounts of D and L isomers of amino acids.(34) A large number of amino acids were found and among them were several which were non-protein in nature. Other evidence for the extraterrestrial nature of the organic matter came from the aliphatic hydrocarbons (which are very similar to those synthesised in prebiotic experiments), the aromatic hydrocarbons (the distribution of which is very similar to a random synthesis from methane or from acetylene) and finally from the $\delta^{13}C$ value of the carbonaceous material which is very different from organic matter in terrestrial sediments.(35) For the first time there has been conclusive proof of the extraterrestrial nature of the amino acids in a carbonaceous chondrite. This analysis has now been extended to other meteorites, the Murray, which fell in Kentucky in 1952 and the Mighei which fell in the Soviet Union in 1866. In every instance the indigenous nature of the amino acids has been established by the fact that equal amounts of the D and L isomers were present, together with a variety of molecules intrinsically nonbiological in nature.(36) Recent expeditions of Japanese and US Polar scientists have found a larger number of meteorites in the Antarctic.(37) Up to the time that these expeditions brought back their samples, the total number of meteorites in collections around the world was only 2000. This number has been dramatically increased by 1300. Eight of these 1300 are carbonaceous chondrites, and two of these eight have recently been examined in our laboratory. Both of them contain amino acids, both protein and nonprotein, with equal amounts of D and L isomers.(38) The Antarctic meteorites have been extremely well preserved, and they are now being handled in the NASA lunar receiving laboratory with the same care and precision with which the lunar samples were handled; the possibility of contamination has thus become minimal. The processes which we have simulated in the laboratory thus seem to have taken place elsewhere in the universe, since the carbonaceous chondrites have preserved for us some samples of the primordial solar nebula.   The chemistry of meteorites is discussed by Brown.(46)

Further dramatic evidence to bolster up the concept of chemical evolution has come to us from the interstellar molecules which the radio astronomers have been discovering over the past few years.(39) Since 1968 many of the molecules that are important to chemical evolution, such as ammonia, water, hydrogen cyanide and carbon monoxide have been found. A recent count gives 50 molecules discovered in the interstellar medium. It is not argued that these molecules contribute directly to the generation of life on some planetary surfaces (see Brown(46)) but conceptually we are pleased to learn that there appears to be an affinity between the hydrogen, carbon, nitrogen and oxygen, giving rise to the very intermediates necessary for the origin of life.(40)

The Voyager Mission to Jupiter has sent back to us information concerning the chemistry of the Jovian atmosphere. Infrared spectra indicate for us the presence of various organic molecules.(41) Laboratory experiments in support of such possibilities have led us to the conclusion that the colours of the planet Jupiter may be a result of organic synthesis.(42) Mixtures of methane and ammonia in varying proportions have been exposed to electrical discharges and to uv light and it is found that as the methane and the ammonia disappear, hydrogen cyanide and acetylene are formed. The analysis of the volatile compounds has also provided us with a wide range of amino-nitriles. It is conceivable that some of these nitriles on hydrolysis will give rise to amino acids. A characteristic result of these reactions has also been the appearance of a red polymer which may have a bearing on the red spot of Jupiter (Fig. 11.4). Spectral analysis in the laboratory may provide some clues by which we might be able to interpret the spectra that the Voyager Programme has provided.

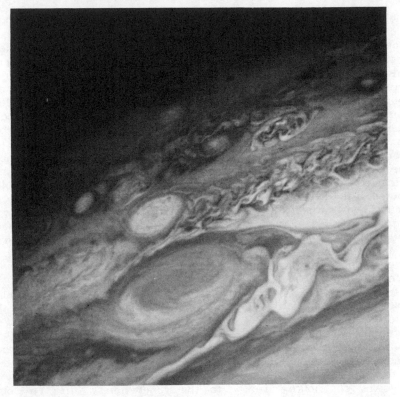

Fig. 11.4. The red spot of Jupiter.

The laboratory experiments, the analysis of meteorites, the interstellar molecules and the current data available from the planet Jupiter lead us to the belief that chemical evolution is truly cosmic in nature and that primordial organic chemistry may be described as organic cosmochemistry.

REFERENCES

(1) Charles Darwin, life and letters, vol. 3, p. 18. (Notes and Memoirs of the Royal Society, London, 1961, 14, No 1.)

(2) A.I. Oparin, 1924; Proiskhozhdenie Zhizni, Izd. Muskovskiy Rabochiy, also RIC Reviews, 1969, 2, 1.

(3) J.B.S. Haldane, 1928, Rationalist, Annual 148.

(4) E. Anders and T. Owen, 1977. Science, 198, 453.

(5) J. Oro, 1963, Ann. NY Acad. Sci., 108, 464.

(6) C. Ponnamperuma and H.P. Klein, 1970, Quarterly Review of Biology, 45, 235.

(7) R.M. Lemmon, 1970, Chem. Reviews, 70, 95.

(8) W.M. Garrison et al., 1951, Science, 114, 416.

(9) S.L. Miller and H.C. Urey, 1959, Science, 130, 245.

(10) C. Ponnamperuma et al., 1969, Adv. Chem., 80, 280.

(11) C. Ponnamperuma, 1978. Proceedings of the Robert A. Welch Foundation Conferences on Chemical Research XXI. Cosmochemistry (W.O. Milligan, ed.) p. 137. Houston, TX.

(12) R. Sanchez et al., 1966, Science, 153, 72.

(13) J. Oro, 1961. Nature Lond., 190, 389.

(14) J. Oro et al., 1959, Arch. Biochem. Biophys., 85, 115.

(15) A. Butlerow, 1861, Ann. 120, 296.

(16) C. Ponnamperuma and R. Mack, 1965, Science, 148, 1221.

(17) C. Ponnamperuma, 1978. In Origin of Life (H. Noda, ed.), p. 67. Japan: Center for Academic Publications.

(18) J. Hulshof and C. Ponnamperuma, 1976, Origins of Life, 7, 197.

(19) A Schimpl et al. 1965, Science, 147, 149.

(20) C. Ponnamperuma and E. Peterson, 1965, Science 147, 1572.

(21) J.C. Sheehan et al., 1967, J. Am. chem. Soc., 78, 1367.

(22) G. Steinman et al., 1965, Science, 147, 1574.

(23) J. Flores and C. Ponnamperuma, 1972, J. Molec. Evol., 2, 9.

(24) A. Gulick, 1955, Am. Sci., 43, 479.

(25) R. Hanel, 1979, private communication.

(26) A. Shimoyama et al., 1978. In Origin of life (H. Noda, ed.), p. 95. Japan: Center for Academic Publications.

(27) A. Negron-Mendoza and C. Ponnamperuma, 1976, Origins of Life, 7, 191.

(28) C. Saxinger and C. Ponnamperuma, 1971, J. Molec. Evol., 1, 63.

(29) C. Saxinger and C. Ponnamperuma, 1974, Origins of Life, 5, 189.

(30) H. Mizutani and C. Ponnamperuma, 1977, Origins of Life, 8, 183.

(31) C. Ponnamperuma et al., Abstr. ACS meeting, Honolulu, Hawaii, April 1979.

(32) M. Schidlowski et al., 1969, Geochim. Cosmochim. Acta 33, 421.

(33) C. Ponnamperuma et al., 1970, Science, 167, 760.

(34) C. Ponnamperuma, 1972, Ann. NY Acad. Sci., 194, 56.

(35) K. Kvenvolden et al., 1970, Nature, Lond., 288, 923.

(36) P. Buhl, 1975. Ph.D. thesis, University of Maryland.

(37) T. Nagata, Proc. Fourth Conf. Antaractic Meteorites, Tokyo, February 21 and 22, 1979 (in press).

(38) C. Ponnamperuma et al., Proc. Fourth Conf. Antarctic Meteorites, Tokyo, February 21 and 22, 1979 (in press).

(39) R.H. Gammon, 1978, Chem. Engng. News, 56, 21.

(40) C. Ponnamperuma, 1973. In Molecules in the galactic environment (M. Gordon and L. Snyder, eds.), New York: Wiley.

(41) R. Hanel, 1979, Science. 204, 972.

(42) C. Ponnamperuma, 1976, Icarus 29, 321.

(43) G. Cairns-Smith, 1979, Chemistry in Britain, 15, 576.

(44) S.P.S., Andrew, 1979, Chemistry in Britain, 1979, 15, 580.

(45) M.D., Brasier, 1979, Chemistry in Britain, 15, 588.

(46) R.D., Brown, 1979, Chemistry in Britain, 15, 570.

DISCUSSION

Siegfried Bauer: There is always a lot of emphasis on the primitive highly reducing atmosphere. What are your ideas about a carbon dioxide dominated atmosphere?

Ponnamperuma: I find it very hard to get away from a reducing atmosphere, at least for some period of time. Maybe it changed very rapidly. It would only be necessary to have a reducing atmosphere for just enough time to make a large abundance of molecules. It's much easier to make those in highly reduced conditions. But, on the other hand, that is not an absolute necessity, because the latest information about atmospheres seems to suggest to us that a lot of material came from outside — the carbonaceous chondrites, or the Anders-Owen idea of comets bringing it. Comets are wonderful. We should go in 1986 and see what is there. That is a marvelous way to verify the concept that a lot of material was brought in by them.

Virginia Trimble: Tell us about Jupiter.

Ponnamperuma: The red spot that you saw — there were two. One was the real one from Voyager. The other was a synthetic one in the laboratory. I think that the methane and ammonia can give you polyamino compounds, and you have chromophores so that the colors can be generated. Up to a few years ago people said, "don't use phosphine (PH_3) because phosphine decays so rapidly". But we have found phosphine in the Jovian atmosphere. If a trace of PH_3 is put into any of these experiments you can get a whole range of colors — the red, the yellow, and so on — but we still have a bit of a problem. We cannot get any meaningful infrared spectra to match with ours. So we are now switching to the ultraviolet, and hopefully something may come out of that.

Michael Hart: In your simulation experiments of primitive atmospheres did you ever get any nucleotides, if so, in what concentrations?

Ponnamperuma: We have not gotten nucleotides directly in that mixture. But we can make the bases, we can make the sugars. We go to the next stage and have to put them together. So, in order to get the nucleotides, we have to go in a stepwise manner. This may be because we have never put phosphorus into these experiments, except for Jupiter. The time has come now to put the PH_3 in and see whether polyphosphates and so on are formed.
     On the side of the protein molecule we can pick out polypeptides. As a matter of fact, in my first electrical discharge experiments we found no amino acids. We got quite a shock — after all, Miller and Urey had done very careful work in 1953. When we examined our soup we found that the amino acids were stuck together. We were fishing out polypeptides of about ten units.

Ben Zuckerman: With the addition of the Antarctic meteorites you now have a fair sample of carbonaceous chondrites. Can you tell how hot a typical one has been heated?

Ponnamperuma: Probably not very hot. Three hundred degrees Kelvin would probably be a maximum temperature. We have now looked at five samples out of thirty-five available. We have theses for graduate students for the next ten years.

12

Chance and the Origin of Life*

Edward Argyle

ABSTRACT

Random chemical reactions in the Earth's primitive hydrosphere
could have generated no more than 200 bits of information,
whereas the first Darwinian organism must have encoded about a
million bits, and therefore could not have arisen by chance. This
information gap is bridged by separating reproduction from
organism, and postulating a reproductive chemical community
that would generate information by proto-Darwinian evolution.
The information content of the initial community could have
been as low as 160 bits, and its evolution might have led to the
first Darwinian cell.

I. INTRODUCTION

It is a widely held view that life will arise spontaneously on the surface
of any planet that provides a suitable physical and chemical environment. This belief is saved from tautology by the generously broad
definitions of 'suitable' that abound in discussions of the origin of life.
Indeed it is almost sufficient to require only that liquid water occur on
the planet's surface, for then it follows that the atmospheric pressure
and ambient temperature will be in ranges that promote a rich variety
of organic reactions.

On the ancient Earth, as today, the simultaneous presence of the
three states of matter along terrestrial shorelines provided reaction
sites and macroscopic transport for most of the planet's chemicals. The
temperature was low enough to confer a substantial lifetime on
thermodynamically improbable molecules formed in sunlit waters, yet
high enough to give speed to the processes of chemistry, and to the

*Reprinted from Origins of Life, Vol. 8, No. 4, 1977, 287-98, with
permission of D. Reidel Publishing Company.

evolution of life. The importance of speed in chemistry and evolution is emphasized by the reflection that a cooler planet than ours, where reaction rates were one fourth as great, would see its sun burn away from the main sequence of stars before it witnessed intelligent life.

This standard scenario of life's origin has been strengthened greatly by the outcome of laboratory experiments in geochemistry (Fox and Dose, 1972). They show that the assumed primitive molecules of our planet's early atmosphere, if supplied with free energy, could form sugars, amino acids, purines, pyrimidines and other life-related organic substances. More recently, the discovery of numerous interstellar organic molecules by radio astroners (Robinson, 1976) has all but confirmed that organic molecules of some complexity are of widespread occurrence in the universe. Hence it is now widely believed that the primordial Earth was the scene of a varied organic chemistry that was rescued from the death of equilibrium by the energy flow from the sun.

At this point in the development of life on Earth the scene goes briefly out of focus. 'Somehow', it is thought, the first very simple, but living reproducing cell was formed. The picture then clears as Darwinian evolution steps in and drives the progeny of that first cell through inexorable stages of experiment and success to a culmination in intelligent life.

It is the purpose of this paper to discuss the events that took place between the settling of the Earth's crust about four thousand million years ago and the appearance of the first living organisms a few hundred million years later. The approach is through information theory, for the light that it throws on the probabilities of random formation of molecules that might have been relevant to the origin of life on Earth.

II. INFORMATION IN SEQUENCES

Information theory has been developed to assist in the design of communication channels and the codes used with them. As such the theory deals with the mathematical properties of a message, or sequence of symbols, and is not concerned with <u>meaning</u>. However, two concessions are made with respect to meaning. If the message is without significance to the receiver, or if it has been received before, it is conceded to contain no information.

The definition of the information content of a message is constructed to meet two intuitive needs. The amount of information should be proportional to message length, and it should increase appropriately with the richness of the alphabet in which it is encoded. The more symbols there are in the alphabet the greater is the number of different messages of length n that can be written in it. The number of permutations of k alphabetic symbols, taken n at a time is

$$N = k^n \qquad (1)$$

where N is the number of different possible messages. The information content, H, of a significant message is defined to be

$$H = \log N. \tag{2}$$

If the logarithm is taken to the base 2 (as always in this paper) the result is expressed in binary units, or bits.

The probability that a random sequence of n symbols would convey the intended message exactly is only $1/N$. In this sense a significant message is an improbable event. If improbability is taken to be the reciprocal of probability, formula (2) states that the amount of information in a message is simply the logarithm of its improbability of occurrence.

An alternative formula for H can be obtained by substituting (1) into (2) yielding

$$\begin{aligned} H &= \log k^n, \\ &= n \log k. \end{aligned} \tag{3}$$

Clearly, the information content of a message is proportional to its length, and also to the logarithm of the number of symbols in the alphabet it uses. In fact, $\log k$ is the information content per symbol. Because (3) is linear in n, the H-values of several messages received in succession (like one longer message) are additive.

The strict validity of these equations requires the condition that each of the k symbols be equally likely to occur anywhere in the sequence. In other words, the code must be non-redundant. This assumption is not seriously violated in the circumstances that will be examined.

That H is a conveniently compact measure of information is seen when formulas (1) and (2) are applied to a biochemical sequence such as messenger ribonucleic acid (mRNA). Assuming that an average gene comprises a string of $n = 1200$ nuclear bases of $k = 4$ kinds, the number of different genes of that length is, by (1), the enormous number

$$N = 4^{1200} = 2^{2400} = 10^{722.5};$$

whereas the information content, using (2), is

$$H = \log 2^{2400} = 2400 \text{ bits,}$$

a manageably small number.

Apart from purposes of illustration it is easier to arrive at this result by using a form of Equation (3). Since $k = 4$, $\log k = 2$ and therefore

$$\begin{aligned} H_{RNA} &= 2n, \\ &= 2400 \text{ bits.} \end{aligned} \tag{3a}$$

There are $k = 20$ different amino acids in a modern peptide chain, or protein. Therefore $\log k = 4.32$ and (3) becomes

$$H_{PEP} = 4.32\,n \tag{3b}$$

for a chain of n amino acids.

An important property of communication channels can now be brought out by comparing the information content of a typical protein with that of the gene that specifies it. Because of the triplet nature of the genetic code the 1200-base gene considered above has 400 codons and therefore is translated into a protein of 400 amino acids. But, by (3b) the information content of the protein is only 4.32 x 400 = 1729 bits. About 671 bits have been lost in translation. This feature of genetics is known as the <u>degeneracy</u> of the genetic code (64 codons represent only 20 amino acids plus start and stop signals). More physically, it is a degeneracy of the translation process. Clearly, a cell equipped with about 60 different transferase molecules (tRNA) instead of 20, could utilize 60 different amino acids for protein manufacture — all without any change in the basic structure of the mRNA. Degeneracy may have been important for the origin of life, and will be mentioned again.

A useful 'benchmark' to the infomation content of living organisms is provided by the common bacterium Escherichia coli (E. coli). Assuming it to have 2500 genes it will encode a total of six million bits. The amounts of information stored in this and other relevant structures are displayed in Table 12.1.

Table 12.1. The Information Content of Various Structures

Structure or sequence	Number of different possible cases or states	Amount of information, bits
A molecule that is certainly present	1	0.
Toggle switch	2	1.
Nuclear base	4	2.
Amino acid	20	4.32
Combination lock	10^6	20.
Modern tRNA molecule	10^{45}	150.
Random experiments	10^{60}	200.
Gene or Protein	$10^{722.5}$	2400.
Virus (50 genes)	$10^{36\ 000}$	1.2×10^5
E. Coli (2500 genes)	$10^{1\ 800\ 000}$	6.0×10^6
Man (100 000 genes)	$10^{72\ 000\ 000}$	2.4×10^8

At this point it is important to make clear the nature of the argument from information theory as it applies to the origin of the first living organism. We are willing, perhaps, to make the optimistic assumption that the formation of random peptide chains of amino acids was thermodynamically favorable, or at least permitted, thanks to the presence of the necessary organic substances, mineral catalysts and available energy. In other words, random peptides were commonplace. But a random chain of n amino acids contains no information unless it has significance for the origin of life. If it, alone amongst such chains, has that significance, its information content will be given by formula (2) with N set to the number of different possible chains of length n. (Equally, depending on one's predilection regarding the manner in which life started, the formation of a significant nucleic acid would generate information.)

If life could have started in but one way it would be possible, in principle, to deduce how it began, given only that there is life. Such an origin would therefore have been, in the language of physics, 'non-degenerate'. But if two or more distinct sequences of amino acids (or bases) could have independently triggered the process, then the origin of life would have been a degenerate event. The mere knowledge that life exists would no longer suffice to deduce the actual mode of origin out of the several that were possible. Clearly, life on earth today is a highly degenerate process in that there are millions of different gene strings (species) that spell the one word, 'life'.

III. THE GENERATION OF INFORMATION

If the first cell was as complex as E. coli, a momentary suggestion made only for illustration, it would have been necessary for our planet's early random chemistry to have generated about six million bits of information by pure chance. The way in which chance experiments can generate information is now examined by looking at a game.

Consider a locked door equipped with a combination lock comprising twenty toggle switches on its outer surface. If all twenty switches are set correctly the door can be opened by grasping and turning the handle. A person having no knowledge of the correct combination can open the door by making a (long) series of random experiments. A few switch positions are changed at random and the handle is tried. This procedure is repeated until the door opens. The experimenter can now write down the combination by looking at the successful switch positions. Clearly, he has acquired 20 bits of information, 1 bit per switch, by making, probably, about one million random experiments. (There are $N = 2^{20}$ $\approx$ one million different possible switch combinations.)

Here again it is seen that a random sequence (of switch positions) is uninformative as long as it is undistinguished. But as soon as one sequence proves to be significantly different from the other million, it carries 20 bits of information. The average amount of information H_r that can be generated by N random experiments is seen to be

$$H_r = \log N, \tag{4}$$

and this is merely a specialization of Equation (2).

To say that random experiments can generate information is perhaps a subjective view not warranted by the facts. It could be argued that the information necessary to open the door is already encoded in the pattern of the electric wiring that connects the switches to the lock mechanism, and that all one need do to acquire that information is to look at the other side of the door! Similarly, it will be true that the sequence of units necessary to the first reproductive structure was already encoded in some deep way in the environment that gave it birth. This may be what the philosophers mean when they say that the potential for life inheres in the very nature of matter.

IV. THE PROBABILITY OF THE ORIGIN

In the standard scenario the Earth's fluids were in a continual turmoil of chemical change that produced organic sequences in a random way. All but one of these long molecules were passive. The unique molecule quickly organized most of the carbon in its environment into copies of itself. The information necessary to reproduce had been transferred from the chemical milieu in which it had lain concealed to a molecule capable of recording it. The random chemistry was then replaced by Darwinian evolution.

Calculation of the probability that a reproductive chemical was formed by chance requires a knowledge of the rate at which relevant random reactions occurred, and the length of time they continued. Neither figure is well established — especially the first — and no realistic calculation of the information-generating power of the environment can be made. On the other hand it is possible to set optimistic upper limits to the reaction rate and the available time, and then to derive the maximum possible value of H_r permitted under the assumptions made.

Assuming that the early waters of Earth contained 10^{44} carbon atoms (Suess, 1975), it is optimistic to take the number of amino acid molecules at 10^{43}. If these molecules were linked in random sequences of average length 10 there would be about 10^{42} peptide chains. If the mean lifetime against extension or breakage for the average chain were 10 milliseconds, and the action continued for 500 million years, the total number of peptides formed would be

$$(10^{42} \text{ peptides}) \times (5 \times 10^8 \text{ yrs}) \times (3 \times 10^7 \text{ s yr}^{-1})$$

$$\times (100 \text{ reactions s}^{-1} \text{ peptide}^{-1}) = 1.5 \times 10^{60}.$$

However, not all of these are different peptides. For the case in which there are 20 kinds of amino acids it can be shown that 98.2% of the 1.5 $\times 10^{60}$ reactions are repeat production of short peptides. Substracting these, there remain

$$N = 2.75 \times 10^{58}$$

different molecules, and the corresponding information content is, by
(4)

$$H_r = 194 \text{ bits.}$$

If the calculations are made for nucleic acids instead of peptides the
result is substantially the same.

It would seem impossible for the prebiotic Earth to have generated
more than about 200 bits of information, an amount that falls short of
the 6 million bits in E. coli by a factor of 30 000. A natural attempt to
save the scenario is to postulate a simpler first cell. However, there is
little to be gained through this proposal. An average virus codes about
2% as much information as E. coli (120 000 bits) and is not capable of
reproducing in an abiotic environment. Rather it must subvert the
metabolic machinery of a regular cell for materials, energy and protein
synthesis. It is difficult to imagine an independently reproductive cell
as simple as a virus (Watson, 1970-1), and even if one can, it helps little
to bridge the enormous information gap between chemistry and life.

Parenthetically, it is interesting to note that if the probability of
the chance appearance of life on Earth seems remote, there is little
comfort to be gained by enlarging the arena to the whole galaxy. Even
if there are 10^9 Earth-like planets in the Milky Way the potential for
random generation of information rises only to 224 bits — less than 0.2%
of the content of the average virus.

Even one gene of average length encodes about 2400 bits, so it is not
useful to speak of a primitive naked gene that reproduced unless it was
so short that it specified a protein of no more than about 33 amino
acids. Whether one prefers to think of the first nucleic acid, the first
gene, the first protein or the first enzyme as the unique structure that
began life, there is the difficulty of visualizing the way so small a
molecule could have commanded the environment to its selective
reproduction.

If life on Earth had a spontaneous origin there must have been an
intermediate mechanism that was capable of augmenting the information content of one or a few early molecules up to the million-bit level
required by the first organism. But before turning to a possible new
mechanism it is worthwhile to consider how Darwinian evolution is able
to generate information so much faster than random experimentation.
The question is philosophically interesting because, as stressed in every
good introduction to the subject, the mutations that provide the raw
material of evolution occur at random and are made without purpose or
goal. Also, the random experiments, mutations, occur very slowly,
about one per generation, and require a great deal of carbon per
individual — about 10^{10} atoms. If most of the ocean's carbon resided in
cells like E. coli, reproducing once per hour for three billion years, the
total number of mutations would be only about 10^{47}, less than 10^{-13} of
the number allowed for random chemical experiments. Nevertheless
Darwinian evolution has been able to amass prodigious amounts of
information in the world's living species. How this was done can be seen
by examining a modified version of the game with the locked door.

V. DARWINIAN GENERATION OF INFORMATION

Consider now a slightly different kind of door – one that opens a millimeter for each switch that is set correctly. The initial switch positions are random. By noticing the door's response, the uninformed operator can open it very quickly. He selects a switch at random and alters its setting. If the door opens a little he makes another experiment. Otherwise he returns the switch to its original (correct) position and then goes on to the next experiment. Twenty systematic trials would have opened the door, but about 60 are necessary if they are random because of needless experimenting with switches already tried.

This door-opening game can be seen to correspond to an extremely simplified model of Darwinian evolution if certain identifications are made. The door is the reproducing organism and its openness is a measure of its success. In each experiment (generation) the door is seen in two states – its original state, and its new state after a switch is thrown. These are the two progeny of a divided cell, one normal and the other mutant. Only the 'fitter' of the two states is tolerated by the operator (the environment), which then destroys the other state (offspring) and leaves the former to serve as original state (parent) for the next experiment (generation).

The 20 bits of information required to open the Darwinian door have been generated in only about 60 experiments (generations), instead of the million required to open the random door treated earlier. Thus

$$H_D = cN, \tag{5}$$

where H_D is the amount of information generated by N Darwinian experiments, and c is a factor that equals $1/3$ in this example.

Of course, it is not accurate to assume that every mutation that makes the base sequence of a gene more like that of a superior gene will necessarily itself be an improved gene. In other words the shortest route to a better gene will not necessarily thread a series of monotonically improving genes. Therefore the formula for H_D gives an upper limit rather than a precise estimate of the amount of information generated in N experiments. Furthermore, the factor $1/3$ is appropriate only to the game described. It will be smaller for longer genes carrying more than 20 bits, for mutation rates less than the 50% used in the game, for larger populations (because of duplicate mutations) and for nearly perfected genes that are harder to improve. Unpublished computer simulations suggest a c-value in the range 10^{-2} to 10^{-6} for plausible simple model organisms.

Even the lower values of c have little impact on the disparity between H_D and H_r for interesting values of N. When $N = 10^{40}$, $H_D = 10^{34}$ and $H_r = 132$. The Darwinian organism acts like a machine for generating information. Its special function is to copy all its genes, including those carrying random alterations to the message. All new messages are then subjected to environmental scrutiny. It is immaterial that only rarely does a new message pass the test. Once approved, it is copied at an exponentially increasing rate and, for a time, becomes the 'standard' message that underlies future attempts to encode even more information about the environment's tolerance for life. This process works because the reproductive power of a population of organisms exceeds the environmental culling that takes place between generations. The cost imposed by genetic experimentation is paid out of surplus reproduction. Meanwhile mutations that are not rejected add information to the genes at the Darwinian rate.

VI. AN INTERMEDIATE MECHANISM

Any proposed new mechanism for generating information faster than possible by pure chance faces a fundamental logical difficulty if it is embodied in an integrated structure. Any structure that reproduces at a rate that outruns decay processes will undergo Darwinian evolution, and for that very reason will be a Darwinian organism. This dilemma can be avoided by dropping the tacit assumption that reproduction is possible only to an integrated structure such as a cell or other living organism.   If an amorphous community of free molecules could reproduce it could also evolve in a proto-Darwinian way, but would not be living because it would contain no structure that independently reproduced itself.

Although a cell such as E. coli would not reproduce in today's environment if it lost its wall there seems to be no impediment to the general idea that the contents of a broken cell (along with the wall fragments), when poured into a benign environment might still carry on the biochemical reactions characteristic of reproduction (Horowitz, quoted in Margulis, 1970). If that happened there would be a multiplication of the populations of biochemicals at the expense of the simpler organic precursors in the region. Such a system would contain no living entity but certain molecules in it would make copies of each other even though no molecule reproduced itself. It must now be asked whether one can postulate a chemically reproductive community containing less than 200 bits of information without forsaking plausibility.

VII. A REPRODUCTIVE CHEMICAL COMMUNITY

An important requirement for plausibility in any proposal for the origin of life is the recalculation of H_r on a less wildly optimistic basis. To this end the assumed concentration of amino acids (or nucleotides) is reduced by the factor 1000 to 10^{40} in all the oceans; the reacting mass is reduced from the entire hydrosphere to those waters shallow enough to permit sunlight to reach the bottom, a factor of 100; and the lifetime of an average molecule is increased to 1000 seconds, a factor of 10^5. The product of these factors is 10^{10} and the lost information entailed by their use is 33 bits. Thus a more plausible value for the amount of randomly generated information is

$$H_r = 161 \text{ bits.}$$

The idea of a reproductive chemical community can be made concrete by adopting the suggestion of Crick et al. (1976) that early protein synthesis took place without the use of even a simplified ribosome. Their scheme achieves translation of a primitive genetic code to protein by means of transfer RNA molecules (tRNA) each carrying a 7-base anticodon loop at its 5' end and a specific amino acid-recognition group at the 3' end. An attempt will now be made to calculate the minimum amount of information required to create the necessary molecules.

In its simplest and most cogent form the new mechanism recognizes the four amino acids glycine, asparagine, serine and aspartic acid, whose modern codons occur in the bottom right hand corner of the codon table. Consequently, four tRNA's are involved in the translation of a mRNA molecule. Because of the degeneracy of the 3-base codon only one bit is required for each base in the mRNA, and therefore also for each of the three variable bases in the 7-base anticodon. The other four fixed anticodon bases are nondegenerate and represent two bits each. Thus 11 bits are required for each of the four 7-base anticodons.

The problem of amino acid recognition is not dealt with by Crick et al. Presumably each tRNA forms a recognition cavity at its 3' end. Allowing 8 bases for the cavity would call for 16 bits for each tRNA. However, the occurrence of considerable degeneracy, especially in primitive structures, is widespread in the biochemical world (Hasegawa, 1975) and it may not be unreasonable to reckon 12 bits per cavity. If functional tRNA's could be this simple they would embody only 23 bits each.

Only one molecule remains – the mRNA, and it will be assigned the remaining information. Subtracting 4 x 23 from the allowed 161 bits leaves 69 bits for a mRNA molecule containing 23 3-bit codons. Without further appeal to degeneracy it is now entirely proper to demand that the mRNA specify the best nonspecific RNA replicating enzyme possible out of all peptides composed of 23 amino acids, each chosen from the set of four.

At this stage in the development of the idea of a reproductive chemical community it would be all too easy to conclude that the replicase would form copies of all 5 RNA's while they were busy making more replicases. Indeed, the formation of the first replicase starts promptly because the principle of additivity of information used in this treatment applies only to the case where all the specified molecules are made at the same time, in the same place and in specified relationship to each other. That is, the 161 bits of information include specification of the correct initial juxtaposition of the 5 molecules to begin translation. But translation will not be completed unless the used tRNA's that drift away from the mRNA after the formation of each peptide bond are brought back again and again by Brownian diffusion until all 22 bonds have been made. Furthermore, all this must happen during the postulated 1000-second lifetime of the molecules. To ensure completion of the first replicase it may be necessary to suppose that the crucial molecules were trapped in a micrometer-sized interstice between particles of clay or other material. After a substantial population of each of the six molecules had been generated the system would become secure against diffusion, and would no longer require, or benefit from, confinement in a small volume.

VIII. PROTO-DARWINIAN EVOLUTION

In the modern cell mRNA codons can be translated into amino acids at the rate of 40 per second, with an error ratio below one in a thousand. Replication of nucleic acids is about equally fast, and nearly a million times more accurate (Watson, 1970-2). In the first reproductive chemical community envisaged here the population would double in every generation if about two reasonably accurate sets of six molecules were produced every 1000 seconds, per mRNA. Thus a high mutation rate combined with sluggish replication and translation would be permitted, as well as expected. If such a system generated information at the rate of 0.1 bit per mutation, ($c = 0.1$ in formula (5)), only a few hours would be required to double the original investment of 161 bits. Thus the onset of reproduction and proto-Darwinian evolution in a randomly generated system that might have taken hundreds of millions of years to arise marks an exquisitely critical point in its history.

In a proto-Darwinian community there are no organisms to compete with each other. The information necessary to reproduction is dispersed in several free RNA molecules. (The central dogma of Darwinian genetics, that information flows from nucleic acid to protein, but not along the reverse path, is already visible in the fact that the peptide molecule, replicase, is not an essential element of the reproductive set.) Tolerable errors in translation or replication lead not to new organisms but to new chemical pathways, and these compete. Because of the slowness of long range molecular diffusion, competition is mostly a local matter. A novel pathway can survive for a time in its own region even though it is not the best in the pool. But in the long run the pool evolves as a whole.

Each mRNA molecule acts as a centre of translation, accepting all appropriate tRNA's that drift into contact with it. Thus each replicase is assembled by random tRNA's in the community surrounding the mRNA. It is noteworthy that the system of Crick et al., as developed here, posits more information in the tRNA's than in the mRNA. If the five RNA's are all regarded as genes it could be said that proto-Darwinian reproduction is hyper-sexual in the sense that nucleic acids mix even more freely than in sexually pairing organisms.

Assuming that the reproductive chemicals eventually leak into new niches and evolve complexity and diversity, there would soon be a variety of specific pathways plied by separate communities of chemicals of ever-increasing molecular weight and specificity. However proto-Darwinian systems lack one important feature of Darwinian populations – safeguards against hybridization. Two divergently evolved chemical communities spilling into a common pool would unavoidably hybridize if their pathways contained any common segments, as would seem very likely. For example, if there had been divergent evolution of genetic codes the indiscriminate mixing of nucleic acids would thwart the interlocking specificities so laboriously built up over their histories and would waste resources prodigiously. This difficulty would dog the path of proto-Darwinian evolution until some means of collecting and sheltering a reproductive set of chemicals was evolved. Perhaps the Darwinian organism was evolution's answer to the problem of hybridization.

It is easy to suggest ways in which a reproductive set of chemicals might clump together in relative isolation from the rest of the community. The difficulty is to have the isolation partial in just the way that permits essential precursor molecules to enter the enclave but prohibits the loss of genetic material until it can be discharged in a self-reproductive clump. That this partial isolation is not easy to specify is just the problem of the origin of life. It probably required a million bits of information, and it is the burden of this paper that that information might have been generated by the proto-Darwinian evolution of a reproductive community that began with less than 200 bits of randomly generated information.

The shortcomings of this scheme to start some kind of rapid information-generating process in the prebiotic soup are too obvious to ignore. The broth is speciously thick and the prospects of the first community are precarious in the extreme. It would be helpful to know more about the catalytic powers of short peptides, especially those containing only two kinds of amino acids.   Such enzymes could be assembled by two tRNA's. If the RNA replicases turn out to be highly degenerate, less information will be needed for their formation. These and other possibilities remain to reduce the improbability of the chance occurrence of a reproducing system.

IX. CONCLUSIONS

Improbable structures can be formed by random trials if the latter are sufficiently numerous. Information theory simplifies the task of separating the possible from the impossible by reducing structural complexity and experimental prodigality alike to a common informational measure expressed in bits.

The calculation of the information-generating power of the Earth's primitive hydrosphere offered here is neither precise nor definitive. Rather it is suggestive that there is an enormous information gap between the products of a random chemistry and the simplest imaginable reproducing organism.

It seem futile to force Darwinian evolution backwards through simpler and simpler organisms to one whose structure could have been the outcome of random trials. Instead it is proposed that special molecules that arose by chance formed a reproductive community of sufficient vigor to start a proto-Darwinian evolution that dominated its development.

Proto-Darwinian evolution will have been significant for the origin of life if at least one reproductive chemical community can be specified by not more than 200 bits of information, and does not lie in an evolutionary cul-de-sac.

If the 200-bit figure is seriously in error, it is too large. If the true figure is less than half this upper limit it will probably be necessary to discover information-generating mechanisms beyond those discussed here. Alternatively, it would be encouraging to discover that enzymes are highly degenerate molecules that economize on information.

Note added in proof: The prospects of the reproductive chemical community described in Section 7 become more promising when it is noted that the system is slightly degenerate. If the specifications of the five molecules are thought of as five words in a 161-bit message it is seen that there will be 5! permutations of word order, and that these do not change the meaning. Moreover, there are 4! possible sets of associations between the anticodons and the amino acid recognition sites of the four tRNA's. Although each such set would require a different mRNA to specify the replicase molecule, the appropriate messenger can always be encoded by the 69 bits allotted to it. Consequently, the postulated system would arise 4! x 5! = 2880 times at diverse places in the hydrosphere. Only one of these systems need have propagative success to start proto-Darwinian evolution.

REFERENCES

Crick, F.H.C., Brenner, S., Klug, A., and Pieczenik, G. (1976). Origins of Life 7, 389.

Fox, S.W. and Dose, K. (1972). Molecular Evolution and the Origin of Life, W.H. Freeman and Co., Chap. 4.

Hasegawa, M. and Yano, T. (1975). Origins of Life 6, 219.

Margulis, L. (1970). Origins of Life, Gordon and Breach, New York, p. 311.

Robinson, B.J. (1976). Proc. Astron. Soc. Aust. 3, 12.

Suess, H.E. (1975). Origins of Life 6, 9.

Watson, J.D. (1970-1). Molecular Biology of the Gene, W.A. Benjamin Inc., New York, p. 503.

Watson, J.D. (1970-2). Molecular Biology of the Gene. W.A. Benjamin, Inc., New York, pp. 297 and 368.

13

Possible Forms of Life in Environments Very Different from the Earth

Robert Shapiro
Gerald Feinberg

ABSTRACT

Speculations concerning extraterrestrial life forms have generally assumed that their basic physical processes will be similar to those employed by life on Earth. For example, they function primarily with carbon compounds, and use water as the fluid medium. In some cases, it has been assumed that the very same biochemicals, such as proteins and nucleic acids, will be used by extraterrestrial life. Such assumptions limit the possible locations suitable for life to planets similar to Earth.

We have considered the requirements for life from a broader point of view and find them to be less restricting. What is required is a flow of free energy, a system of matter capable of interacting with the energy and using it to become ordered, and enough time to build up the complexity of order that is associated with life. These conditions may be met in a variety of environments, some very different from Earth. They may give rise to life forms using solvents other than water, chemical systems making little use of carbon, and even forms based on physical interactions, rather than molecular transformations. It is possible that life is very prevalent throughout the Universe.

INTRODUCTION

We can make only a brief presentation here of our ideas on the extent of life in the Universe. A much more detailed account is given in our book: "Life Beyond Earth: The Intelligent Earthling's Guide to Life in the Universe."(1)

In discussions concerning intelligent life elsewhere, the assumption is often made that such life will develop only in circumstances

113

resembling those on Earth. Estimates are then given of the number of Earthlike planets in our galaxy, as suitable locations in which life might arise. Such estimates may vary, according to the pessimism or optimism of the observer, from the very few(2) to a billion or more.(3) Only those planets are considered habitable which fall into a limited zone around each star. In that zone, liquid water can be present on the surface, and carbon compounds will be abundant. If this view were correct, even in the most optimistic form, then life would be a rare phenomenon, confined to only an insignificant fraction of the material in the Universe. From an extreme pessimists' viewpoint, as expressed else-where in this conference, life may have originated only on the planet Earth. The idea of the specialness of the Earth is of course an old one, and has been expressed many times in theology.

We represent a very different point of view: that the generation of life is an innate property of matter. It can take place in a wide variety of environments very different from the Earth. The life forms that evolve will also be very different from those familiar to us, in harmony with the conditions present there.

Because the point of view opposite to our own is widespread, we think it is important to summarize the reasons presented for it. The most obvious one is the basic unity of Earthlife, the only type of life we have encountered. All living things that we are aware of use the same basic set of chemicals to organize their metabolism, with vital roles for proteins and nucleic acids. A person in a Chinese village might similarly be convinced that only the Chinese language existed, as it was the only one spoken by all the people he knew. In fact, no firm conclusion can be drawn at all from cases where only a single example of a phenomenon is at hand, and a good strategy would be to search for additional examples.

Other arguments have been made on the basis of scientific princi-ples, and we have found that their adherents could be grouped into two categories, which we have called "predestinist" and "carbaquist."

THE PREDESTINISTS

The predestinists hold that the limited set of chemicals that charac-terize our biochemistry will be the inevitable result of random chemical synthesis, throughout the universe. We can express this in direct quotes: "The implications of these results is that organic syntheses in the Universe have a direction that favors the production of amino acids, purines, pyrimidines and sugars: the building blocks of proteins and nucleic acids. Taken in conjunction with the cosmic abundance of the light elements, this suggests that life everywhere will be based not only on carbon chemistry, but on carbon chemistry similar to (although not necessarily identical with) our own."(4) "The essential building blocks of life — amino acids, nitrogen-bearing heterocycle compounds and poly-saccharides — are formed in space. These compounds occur in large quantitites throughout the galaxy."(5)

These concepts are imaginative but unfortunately not supported by the facts. The presence of organic compounds in interstellar dust clouds has been demonstrated by spectroscopy.(6) However, the molecules that have been identified definitely are small in size, and do not include those considered to be the essential building blocks of Earthlife. Meteorite analyses provide another source of information about extra-terrestrial organic chemistry.(7) Amino acids are present there, but as part of a complex mixture containing many compounds irrelevant to our biochemistry. There is no preference for the synthesis of compounds important to Earthlife. Finally, we can consider prebiotic simulation experiments of the kind initiated by Stanley Miller and Harold Urey.(8) Good yields of amino acids have frequently been obtained. Although there is considerable doubt that the conditions used do represent those of the early Earth,(9) such experiments do demonstrate that the preparation of amino acids from very simple compounds is feasible. They do not, however, show that the result is inevitable. In fact, the very first effort by Stanley Miller was unsuccessful: "the next morning there was a thin layer of hydrocarbon on the surface of the water, and after several days, the hydrocarbon layer was somewhat thicker."(10) No amino acids were detected at all.

The course of organic synthesis in the universe may on some occasions turn in the direction of the chemicals of Earthlife, but it is clear that it can take other directions as well. We are not the inevitable predestined endpoint of cosmic evolution.

THE CARBAQUISTS

A somewhat different point of view is taken by the carbaquists. They feel that life elsewhere in the universe must resemble that present on Earth, because no other basis for life can exist. Our own chemical system, particularly in its use of water, as a solvent, and carbon, as the key building block of large molecules, is uniquely fit for the purpose of sustaining life. The notion of the fitness of our environment was advocated early in this century by the American chemist Lawrence Henderson,(11) and has had a number of advocates more recently. Again, we will quote them directly.

I have become convinced that life everywhere must be based primarily upon carbon, hydrogen, nitrogen and oxygen, upon an organic chemistry therefore much as on the Earth, and that it can arise only in an environment rich in water.(12)

. . . so I tell my students: learn your biochemistry here and you will be able to pass examinations on Arcturus.(13)

The capacity for generating, storing, replicating and utilizing large amounts of information implies an underlying molecular complexity that is known only among compounds of carbon.(4)

Water _does_ have certain special properties as a solvent. One is the greater amount of heat that is needed to melt, warm up, or boil a quantity of water, as compared to the heat needed for most other solvents. Bodies of water thus tend to stabilize the climate of their environment. This property may be pleasant, but it is hardly essential to life. Other physical features could work to stabilize a climate. Alternatively, living beings could adapt by many strategies, such as spore formation or migration, to changes in temperature.

Another property of water that is greatly admired by the carbaquists is its polar character. Water is a good solvent for charged substances, and an appropriate medium for the conduct of transformations of such substances. However, an enormous number of reactions have been described by chemists which take place in less polar, or nonpolar solvents. The absence of water is in some cases essential to the course of the reaction. There is no reason why such reactions could not be the basis of a metabolic scheme to sustain life.

The special properties of carbon include its ability to bond to itself in long chains, and to form bonds to four other atoms at one time. An enormous number of compounds containing carbon can therefore exist. It is not necessary for an atom to bond to itself to form long chains, however; the chains could be made of two or more atoms in alternation. Furthermore, it is conceivable that a basis for life could be constructed using an alternative chemistry in which the possibilities were not as vast as those of carbon. The situation can be compared with the problem of information storage in printed form. The English language uses twenty-six capital letters, twenty-six small letters, space, and a number of punctuation marks to do this: about sixty characters in all. The same amount of information can be stored by a computer, using only the two symbols 1 and 0. Six lines must be used to hold the contents of one English line, but the data is stored just the same. In the same way, a less complex chemistry could serve as the genetic basis for life, perhaps with a larger number of components needed in each molecule, cell or other unit.

The carbaquist viewpoint cannot be made convincing by the type of reasoning its adherents have presented, but it also cannot be refuted strictly on the basis of reason and analogy. We can best proceed by searching for alternate life forms. The discovery of one, with a different physical or chemical basis, would quickly settle the issue. Failure to do so, after a number of extraterrestrial life forms had been encountered, would move us toward acceptance of the carbaquist argument.

A DEFINITION, AND SOME CONDITIONS

In seeking a broader view of the possibilities of life in the universe than that put forward by the carbaquists and predestinists, we have started by framing a definition of life that is independent of the local characteristics of Earthlife:

Life is the activity of a biosphere. A biosphere is a highly ordered system of matter and energy characterized by complex cycles that maintain or gradually increase the order of the system through an exchange of energy with its environment.

An important feature in our definition is the identification of the biosphere as the unit of life. The history of life on Earth then becomes the tale of the continuous survival and evolution of the biosphere from its origin on the prebiotic Earth. Replication, and subdivision into organisms and species have been strategies adopted by our own biosphere to ensure its own survival but they need not be the methods used by an extraterrestrial biosphere.

The association of life with increasing order, and the emphasis on the need of a flow of energy to sustain this, are related to the ideas of the physicists Erwin Schrödinger(14) and Harold Morowitz.(15) Our biosphere has a number of easily recognizable forms of order, all in a very high degree. It contains very large numbers of a few thousand small organic molecules and none (or very few) of millions of other molecules. The presence of only a few distinct types of large molecules in living things is a second type of order. An additional dramatic type is the near identity of the base sequences in the DNA in different cells in a multicelled organism. If comparable amounts of order (though perhaps of a very different type) were to be found in an extraterrestrial environment, this would be a strong sign of the presence of life.

With the definition in hand, we can now make a list of several conditions that must be met if life is to originate and develop in a particular location.

1. A flow of free energy: Any of a number of different types could suffice, such as light energy and chemical energy, as on Earth, but also other forms of electromagnetic radiation, such as infrared light and X-rays. Other forms of energy such as streams of charged particles, heat differentials, and nuclear energy could also be used.

2. A system of matter capable of interacting with the energy and using it to become ordered: The nucleic acids and proteins of Earth need not be the basis of this order. In fact, it needn't depend on chemical reactions at all but could be based on physical processes such as the movement of particles, or molecular rotations. However, some systems will be superior to others. A liquid or a dense gas is preferable to a solid for the conduct of reactions. Helium is a poor choice as the base for the development of an ordered system based on a multiplicity of chemical compounds.

3. Enough time to build up the complexity that we associate with life: This is a critical question that determines the scope of life in the Universe. The rate of a process, such as the chemical reactions between molecules widely scattered in outer space, may be so slow that the entire lifetime of the Universe to date has been insufficient for appreciable order to develop. In another case, a drastic change in an environment, for example, the conversion of a star to a supernova, could destroy the base upon which order has been accumulating.

ALTERNATIVE BASES FOR CHEMICAL LIFE

With the requirements in mind, we have tried to generate some speculative suggestions for life forms that would function on a chemical basis different from that on Earth.

1. Life in ammonia: Liquid ammonia, perhaps with some water, would serve as the solvent. This could occur on a planet with temperatures near $-50^{\circ}C$. Weaker chemical bonds, such as nitrogen, would predominate in metabolism.

2. Life in hydrocarbons: A mixture of hydrocarbons would function as the solvent. A wide range of temperatures could be accommodated, depending on the composition of the mixture. Processes involving charged species would play only a small role. Reductive reactions, such as hydrogenation, could be used as an energy source.

3. Silicate life: Silicates exhibit a rich chemistry with a propensity for forming chains, rings and sheets, as in minerals on Earth. At a temperature above 1000°, the medium would become liquid, and could serve as a basis for evolving chemical order. This process could occur on a planet quite close to its sun, or in the molten interior of a planet, such as Earth.

ALTERNATIVE CHEMICAL LIFE WITHIN OUR SOLAR SYSTEM

Locations within our own solar system offer the best opportunities for the detection of alternative life forms in the immediate future. Some of the more promising possibilities are listed below:

1. Earth: in interior magma, or within specialized niches on the surface.
2. Mars: if life is present, it is probably based on carbon and water.
3. Jupiter: many possibilities exist in varied environments.
4. Ganymede, Callisto: life within water oceans beneath the crust.
5. Io, Venus: life in liquid sulfur.
6. Titan: ammonia or hydrocarbon based life.

PHYSICAL LIFE

As we suggested earlier, many physical processes may serve for the storage and increase of order. We have called such systems "physical life," and list some speculative possibilities below.

1. Plasma life within stars: Such life would be based upon the reciprocal influence of patterns of magnetic force and the ordered motion of charged particles. It could exist within our own Sun. Individual creatures are called "plasmobes."

2. Life in solid hydrogen: This could occur on a planet with a temperature of only a few degrees Kelvin. Infrared energy would be absorbed and stored in the special arrangement of ortho- and para-hydrogen molecules.

3. Radiant life: Life would be based upon the ordered patterns of radiation emitted by isolated atoms and molecules in a dense interstellar cloud. Individual beings, called "radiobes" could exist.

It may be difficult to think of such systems as being alive, with the capability in some cases of developing organisms, complex ecologies and even civilization. We must remember that the association of a protein with a nucleic acid, when viewed abstractly, also does little to convey the wonders, such as elephants and Sequoia trees, that ultimately arise from it.

CONCLUSIONS

Our examples have not been presented in order to make specific predictions, but rather to suggest the vast variety of forms that life elsewhere may take. If we were gifted with a vision of the whole Universe of life, we would not see it as a desert, sparsely populated with identical plants which can survive only in rare specialized niches, but rather as a botanical garden with countless species, each thriving in its own setting. This garden awaits our exploration.

REFERENCES

(1) Feinberg, G. and Shapiro, R. (1980). Life Beyond Earth. The Intelligent Earthling's Guide to Life in the Universe, William Morrow Co., New York.

(2) Hart, M.H. (1979). Icarus 37, 351.

(3) von Hoerner, S. (1978). Die Naturwissenschaften 65, 553.

(4) Horowitz, N.H. (1976). Accounts of Chem. Research 9, 1.

(5) Hoyle, F. and Wickramasinghe (1977). New Scientist (Nov. 17), 174.

(6) Gammon, R.H. (1978). Chemical and Engineering News (Oct. 2), 21.

(7) Lawless, J.G., Folsome, E.E. and Kwenvold, K.A. (1972). Scientific American 226, 38.

(8) Miller, S.L. and Urey, H.C. (1959). Science 130, 245.

(9) Kerr, R.A. (1980). Science 210, 42.

(10) Miller, S.L. (1974). In The Heritage of Copernicus: Theories Pleasing to the Mind (editor: J. Neyman), MIT Press, Cambridge, MA, 228.

(11) Henderson, L.J. (1913). The Fitness of the Environment, Macmillan Co., New York.

(12) Wald, G. (1974). In Cosmological Evolution and the Origins of Life, (editors: J. Oro et al.) Vol. I, D. Reidel, Dordrecht, Holland, 7.

(13) Wald, G. (1973). In Life Beyond Earth and the Mind of Man, (editor: R. Berendzen) NASA Scientific and Technical Information Office, Washington, D.C., 15.

(14) Schrodinger, E. (1956). What Is Life?, Anchor Books, New York.

(15) Morowitz, H. (1968). Energy Flow in Biology, Academic Press, New York.

DISCUSSION

Eric Jones: Concerning dense interstellar clouds, they have very short lifetimes. They are blown apart by hot stars and by supernovae.

Feinberg: Well, we looked at that a bit. The lifetime depends somewhat on the density. The collapse times can be reasonably long. At a density of 10^4 atoms per cubic centimeter the collapse times are millions of years.

Freeman Dyson: There is a new book by Robert Forward on life on neutron stars. It's a good story.

Feinberg: We have also speculated on life on a neutron star using Ruderman's polymeric atom which, to me at least, sounds eerily similar to nucleic acid polymers.

Cyril Ponnamperuma: I congratulate Drs. Feinberg and Shapiro for discussing life that we do not know about. We have enough trouble understanding life that we know about. How would you detect life of the kinds that you describe? Do you have one test to find out? There may be life lurking around here.

Feinberg: I can't give you an answer that would send NASA out. However, there is certainly none of that kind of life here in the room. That's easy to tell by just making a spectral analysis of the light.
   One of the consequences of the things we said is that in looking for life, leaving apart intelligent life as a special case, we should look for biospheres. We should not try looking specifically for individual living things: (a) because the biospheres are the unit of life and (b) if you think of the Earth, the way of detecting life on the Earth from a distance, until 50 years ago, was not from what human beings did but from what the blue-green algae and the denitrifying bacteria did. We think that

the same principle also applies in other cases.   When we look for extraterrestrial life we should structure the searches in terms of looking for biospheres rather than in looking for individual creatures. How you look for biospheres depends on which biosphere you are after, but that is not something that I can describe in a few words.

If you believe that life is a consequence of the Schrodinger equation for 10^{50} atoms, which I guess I believe, then there isn't anything mysterious about it. You know what the fundamental equations are and they lead eventually to life. The mysteriousness comes about because we are stupid. We don't see that that equation gives you that consequence, and that that's a consequence that you always have. Being surprised is a human condition, and you can be more or less surprised depending on how smart you are.

14

Cosmology and Life in the Universe*

J. Richard Gott, III

ABSTRACT

The detectability of civilizations at cosmological distances is investigated. For communications at cosmological distances frequency standards tied to the $T = 2.7^{\circ}K$ cosmic microwave background are proposed: 28 GHz ($h\nu = 1/2\ kT$) and 56 GHz ($h\nu = kT$). Modest searches from the ground at 28 GHz could be carried out with present technology. These would provide an important supplement to searches for life in our galaxy alone.

I. INTRODUCTION

In this paper we will examine the possibilities for detecting life at cosmological distances. In section II we will estimate the number of habitable planets in the observable universe. In section III we will estimate the number of civilizations visible to us as a function of the probability of life developing and whether colonization occurs. In Section IV we discuss how civilizations at cosmological distances can be detected using frequency standards tied to the cosmic microwave background radiation.

II. THE NUMBER OF HABITABLE PLANETS

To calculate the number of habitable planets in the observable universe we must determine the local density of habitable planets and use the appropriate cosmological model.

---

*The author wishes to acknowledge support from the Alfred P. Sloan Foundation.

The discovery of the $2.7^{\circ}K$ cosmic microwave background by Penzias and Wilson (1965) has led to virtually unanimous acceptance of the big bang model for the origin of the universe. The simplest big bang models are the Friedmann models with zero cosmological constant which represent solutions to Einstein's field equations. These models are completely characterized by their values of H_0 and Ω. Hubble's constant H_0 measures the current expansion rate of the universe. Current estimates of H_0 fall in the range 50 to 100 km s^{-1} Mpc^{-1}. For convenience throughout this paper we shall adopt $H_0 = 50$ km s^{-1} Mpc^{-1} (cf. Sandage and Tammann 1975). The quantity $\Omega = 8\pi G\rho/3H_0^2$ measures the ratio of the current density to the critical density required to eventually halt the expansion of the universe. If $\Omega > 1$ the universe has positive curvature, is closed with finite volume, and at some point in the future, the expansion of the universe will halt and the universe will contract again to a dense state. If $\Omega = 1$, the universe is spatially flat (Euclidean) with infinite volume and will expand forever. If $\Omega < 1$ the universe is negatively curved, has infinite volume and will expand forever. We will consider all three types of cosmology.

The number of habitable (earthlike) planets in our galaxy is estimated to be of order 10^9 (Bracewell 1975). The total blue luminosity of our galaxy is of order $L = 1.5 \times 10^{10} L_\odot$. ($L_\odot$ = luminosity of the sun). Gott and Turner (1976) estimate the mean blue luminosity density of the universe to be $\rho_L \sim 4.7 \times 10^7 L_\odot$ Mpc^{-3}. Combining these estimates gives a mean cosmological density for habitable planets of 3×10^6 (HP) Mpc^{-3}.

Consider an $\Omega > 1$ cosmology. This universe has the geometry of a three-sphere. As an example take $\Omega = 2$. Then the radius of curvature of the universe is $a_0 = cH_0^{-1} = 6000$ Mpc and it currently has a total finite volume of $V = 2\pi^2 a_0^3 = 4.3 \times 10^{12}$ Mpc^{-3} (cf. Weinberg 1972). The number of habitable planets in the universe is $\rho_{HP} V = 10^{19}$. There will be other intelligent civilizations in the universe so long as the probability for a habitable planet to spontaneously develop intelligent life is greater than 10^{-19}.

If $\Omega \leq 1$ the universe has an infinite number of habitable planets and as long as there is a finite probability of forming intelligent life on each habitable planet there will be an infinite number of intelligent civilizations in the universe. If the probability is very small, however, the nearest of these civilizations will be very far away.

The above models are those with the simplest topologies. Einstein's field equations tell us about the geometry of spacetime but not its topology. For example, a space-like slice of the $\Omega = 1$ cosmology is an infinite Euclidean three-dimensional space. We can construct a cube with side L in this space and topologically identify the top with the bottom, the front with the back, and the left with the right side. In this way we can make an $\Omega = 1$ universe with a T^3 toroidal topology and a finite volume $V = L^3$. Lower limits on L can be set by our failure to find multiple images of clusters of galaxies. Gott (1980) finds L must be > 400 Mpc. Thus if $\Omega = 1$ there must be at least $\rho_{HP} V = 2 \times 10^{14}$ habitable planets in the universe (and an infinite number if the topology is simple). There are also multiply connected $\Omega < 1$ cosmologies. These all have $V > 0.98 \, a_0{}^3$ where $a_0 \geq 6000$ Mpc is the radius of curvature of the negatively curved space (cf. Gott 1980). Thus the $\Omega < 1$ cosmologies must have at least 6×10^{17} habitable planets (and an infinite number if the topology is simple).

Because of the finite age of the universe we can not see all of it by the present epoch. The limit of observation is called the particle horizon. At the present time we can see everything within the horizon radius. Photons from beyond the horizon radius have simply not had time to reach us by the present epoch. Big bang models have ages that are comparable with $H_0{}^{-1}$ and so a crude estimate of the number of habitable planets within the horizon is $4 \pi \rho_{HP} r_H{}^3 / 3 = 3 \times 10^{18}$ where $r_H \sim c H_0{}^{-1} = 6000$ Mpc. To compute an exact estimate one must use the curved spacetime metrics of the appropriate cosmological model. We have performed this calculation for three representative cases using the standard Friedmann metrics (cf. Weinberg 1972). The number of habitable planets within the horizon is 6×10^{18} for $\Omega = 2$, 2×10^{19} for $\Omega = 1$, and 2×10^{21} for $\Omega = 0.1$. These numbers are not many orders of magnitude away from the crude estimate. Even if the universe is infinite we only see a finite portion of it at the present epoch. If the probability of forming intelligent life on a habitable planet is less than 5×10^{-22} then we are unique within the observable universe.

As time goes on the horizon radius increases and we see more and more of the universe. Dyson (1979) has shown that in an $\Omega < 1$ cosmology it would not violate thermodynamics for a civilization to maintain itself indefinitely. By operating at lower and lower temperatures as the universe expands the civilization can, in principle, produce an infinite number of thoughts in an infinite time if $\Omega < 1$. (although insurmountable practical problems may be encountered). Given enough time any two civilizations should come within each other's horizon. But Dyson points out that communication between civilizations will be very inefficient if their separation is much greater than $10 \, a_0$ (the radius of curvature) unless there are intermediate relay stations. The volume within a radius of $10 \, a_0$ in the negatively curved space is $V = a_0{}^3 \, \pi \, (\sinh (20) - 20)$ (Gott, 1980). The number of habitable planets within this radius is $\rho_{HP} V = 5 \times 10^{26}$. If $\Omega < 1$ and the probability of forming intelligent life is many orders of magnitude less than 10^{-27} per habitable planet then civilizations can not only not communicate with each other at present but will have some difficulty communicating directly with each other even in the far future.

III. THE NUMBER OF OBSERVABLE CIVILIZATIONS

If Drake and Sagan are correct that the probability of forming intelligent life on a habitable planet is large then there are many intelligent civilizations in our galaxy. (cf. Shklovskii and Sagan 1966 for discussion). Proponents of this scenario would argue that interstellar travel is difficult and colonization is not prevalent so that even though there are many civilizations in the galaxy it is not surprising that we have not been colonized. Hart (1975) and others have stressed the role of colonization. They claim that it is likely for civilizations to embark on interstellar colonization and that if the earth were not the first civilization in the galaxy the solar system would have been colonized long ago. Since this appears not to have occurred, Hart argues that we must at least be unique within our galaxy; so less than one in 10^9 of the estimated habitable planets in the galaxy gives rise to intelligent life by the present epoch. We will consider both colonizing and non-colonizing models.

In its most extreme form the colonization hypothesis would predict that once a civilization formed it would colonize every habitable planet within a sphere expanding with a colonization velocity V_c. The most optimistic estimates take V_c as large as 0.1 c (cf. Hart 1975). If we are typical and have not been colonized, the colonization spheres must not have completely filled space by the present epoch. For purposes of discussion assume that the earliest civilizations began appearing ~ 2 billion years ago. We could detect signals from civilizations up to 2 billion light years away or ~600 Mpc. This is small relative to the radius of curvature of the universe for $\Omega \leq 1$ and $2 > \Omega > 1$ where $a_0 \gtrsim 6000$ Mpc. Since Δt = 2 billion years is short compared to the age of the universe all the cosmological solutions approximate Euclidean results. The volume in which civilizations may be detected is thus of order V ~$4\pi(600$ Mpc$)^3/3$ and contains ~3.0×10^{15} habitable planets. An exact calculation for the $\Omega = 1$ case gives a number only 27 percent larger than the simple calculation above. The number of habitable planets within 2 billion light years is much less than the number within the horizon. So there may be galaxies within the horizon which will develop civilizations by the present epoch but where the look-back time is so large that the civilization has not developed by the epoch at which we see it. Thus if the probability of a habitable planet forming intelligent life is less than 10^{-16} we probably will not detect any other civilizations.   With maximal colonization the oldest civilizations will have colonization spheres of radius $V_c \Delta t$~60 Mpc, the median civilization will have a colonization sphere of ~30 Mpc. Thus the median number of habitable planets in a mega-civilization will be of order $3 \times 10^6 \times 4\pi(30)^3/3 = 3 \times 10^{11}$. The sizes of the mega-civilizations are just large enough that they begin to be approximately fair samples of the universe and we can calculate using mean densities and not worry about the clustering of galaxies. The amplitude of the clustering is less than unity on scales larger than 10 Mpc (Peebles 1974). The median angular radius of a mega-civilization as seen from the earth is

$$\theta_m \sim \sin^{-1}((2^{1/3}-1)V_c/c) \sim 1.5^\circ \tag{1}$$

This assumes $\Delta t \ll H_0^{-1}$ and includes the effects of look-back time. If less than half of the planets in the universe are colonized by the present epoch (earth not yet colonized) the number density of mega-civilizations must be less than ~ $(2\overline{V})^{-1}$ where $\overline{V} = \pi V_c^3 (\Delta t)^3/3$ is the mean comoving volume at t_o of mega-civilizations formed randomly between $t_o - \Delta t$ and t_o. Thus the number density of mega-civilizations is less than 2×10^{-6} Mpc^{-3}. The number of inhabited planets visible from earth $\leq [\rho_H P/70] \cdot [4\pi(c\Delta t)^3/3] \sim 4 \times 10^{13}$. Thus if colonization is rampant the probability of life evolving spontaneously on a habitable planet can be as low as 7×10^{-13} and still produce 4×10^{13} inhabited planets visible to us. (Both of the calculations above assume Δt is small compared to the Hubble time (H_o^{-1}) and include the effects of look-back time)., There will be at most ≤ 500 mega-civilizations visible at distances up to 2 billion light years. The number of mega-civilizations in a radial interval dr is proportional to $4\pi r^2$ (1-r/cΔt) dr for $r < c\Delta t$. The factor in parenthesis is due to the fact that at large distances we are looking back to a time where the number of mega-civilizations that have formed is smaller. Because of this effect we see more mega-civilizations nearby and in fact half of the observed mega-civilizations lie at distances of less than 1.2 billion light years. Furthermore, the nearby mega-civilizations will on average have larger colonization spheres and a larger number of inhabited planets. (Because the look-back time to the nearby mega-civilizations is short they are older when we see them and have had longer to colonize.) Half the visible inhabited planets lie within 0.7 billion light years. The above can be considered as the most optimistic colonization model which leaves an appreciable probability that the earth has not yet been colonized.

If the probability of life evolving spontaneously on a habitable planet is only 1.5×10^{-15} then only 1 mega-civilization will be visible. The median value for its distance is 1.2 billion light years. The median value of its age at the epoch we are observing is 0.4 billion years and it has a colonization sphere with radius of 12 Mpc. With a uniform density cosmological model this would imply 2×10^{10} civilizations. (In a low Ω cosmological model the density 1.2 billion years ago is 1.2 times the present density. Thus the estimate 2×10^{10} should be increased by 20 percent. In general we will ignore corrections of this magnitude, i.e., first order in $\Delta t/H_o^{-1}$.) Galaxy clustering will increase this estimate somewhat. The covariance function $\xi(r)$ measures the excess probability of finding a galaxy at a distance r from a galaxy chosen at random. (cf. Peebles 1974.) Recent estimates give $\xi(r) = 40$ (r/Mpc)$^{-1.8}$.

$\xi(r)$ measures the ratio of the mean excess density at radius r to the mean density of the universe. The covariance function at these scales is produced by bound systems whose density is not changing; 1.2 billion years ago ρ was 1.2 times the present density so the amplitude of $\xi(r)$ at that epoch was lower by the same factor; $\xi(r) \sim 33$ (r/Mpc)$^{-1.8}$. The number of civilizations in a colonization sphere of radius R_o is given by:

$$n = \rho_{HP} \int_o^{R_o} 4\pi r^2 \ (1 + \xi(r)) \ dr \qquad (2)$$

$$n = \rho_{HP} \cdot \frac{4\pi R_o^3}{3} \left(1 + \frac{99}{1.2} \left(\frac{R_o}{Mpc}\right)^{-1.8}\right) \qquad (3)$$

For R_o = 12 Mpc the extra correction factor is 94 percent. Combining the 20 percent correction due to the higher density of the universe and the 94 percent clustering correction we find approximately 5×10^{10} civilizations in the mega-civilization.

If the probability of intelligent life developing spontaneously on a habitable planet is in the range 10^{-9} to 1 then each galaxy is expected to have at least one civilization and we must assume that no colonization takes place (otherwise the earth would have been colonized). We will assume that civilizations form uniformly in the interval from 2 billion years ago to the present and that they transmit continuously after their formation. With probabilities in this range our galaxy will have from 1 to 10^9 civilizations and we will see outside our galaxy a total of 7×10^5 to 7×10^{14} civilizations. In this model our galaxy must be typical and the extragalactic civilizations always outnumber the civilizations in our galaxy by a factor of 7×10^5. The median distance of these civilizations is 1.2 billion light years.

For probabilities in the range 7×10^{-13} to 10^{-9} colonization at $V_c = 0.1$ c is also not allowed. We expect no other civilizations in our galaxy but a total of 5.0×10^2 to 7×10^5 civilizations overall with median distance 1.2 billion light years. If the probability of intelligent life developing on a habitable planet is less than 1.5×10^{-15} we will find no civilizations at all.

In all models where civilizations are visible the vast majority are extragalactic. The results are summarized in Table 14.1.

Table 14.1 assumes that civilizations transmit continuously after formation. If civilizations transmit only a fraction f of that time then the number of visible civilizations will be multiplied by f.

If each civilization lives for a time L_t after formation (during which time it transmits) and then it dies, and L_t is short compared with a billion years then the results are affected as follows. The colonizing models are unchanged because if L_t is long enough for each civilization to send out colonies then a planet with a dead civilization will be recolonized after a short period. The entries in Table 14.1 where colonization is not allowed ($7 \times 10^{-13} < P_i < 1$) are multiplied by the following factors: The number of other civilizations visible in the galaxy is multiplied by $L_t / (2 \times 10^9 \text{ yrs})$. The number of civilizations visible is multiplied by $4L_t / (2 \times 10^9 \text{ yrs})$. The additional factor of 4 in the second case is due to the fact that in this model the density of civilizations from two billion years ago to the present is constant rather than increasing with time. The median distance of visible civilizations is 1.6 billion light years rather than 1.2 billion light years.

The observed luminosity density of the universe is proportional to H_0 so if H_0 were increased from 50 to 100 km s^{-1} Mpc^{-1} the number of civilizations visible would be doubled.

IV. COSMIC FREQUENCY STANDARDS

For communication within the galaxy the 21-cm line proposed by Cocconi and Morrison (1959) is promising. All the observers can agree on a velocity standard (such as the velocity of the center of the galaxy)

Table 14.1. Detectability of Civilizations

Probability of Intelligent Life Forming Spontaneously on a Habitable Planet	Colonization Allowed	Number of other Civilizations in our Galaxy	Number of Civilizations Visible	Median Distance 10^3 lt. yr.
1 to 10^{-9}	No	10^9 to 1	7×10^{14} to 7×10^5	1.2
10^{-9} to 7×10^{-13}	No	0	7×10^5 to 5×10^2	1.2
7×10^{-13} to 1.5×10^{-15}	Yes	0	4×10^{13} to 5×10^{10}	0.7 to 1.2
$<1.5 \times 10^{-15}$	Yes	0	0	None

and tune their receivers to that standard. Alternatively signals can be sent to specific target stars so that the signal will have the 21-cm line frequency in the rest frame of the target star. Several unsuccessful 21-cm searches have been undertaken so far. To look for life beyond our own galaxy we must find cosmological frequency standards for communication.

Frequency standards based on line radiation such as the 21-cm line are not useful for intergalactic communication because of the cosmological redshift. In principle one could identify the 21-cm rest wavelength in each galaxy to be studied (redshift could be measured to ~10 km s^{-1} giving a frequency standard of $\Delta\nu/\nu = 3 \times 10^{-5}$). But this would be very time consuming for a survey out to 600 Mpc.

Fortunately the cosmic black body radiation discovered by Penzias and Wilson (1965) provides a frequency standard with the desired properties. It has a Planck spectrum:

$$I(\nu)d_\nu = \frac{2\pi h\nu^3}{c^2} \frac{d\nu}{\exp(h\nu/kT)-1} \tag{4}$$

with a measured characteristic temperature $T = 2.7°K$. This spectrum is consistent with the observations over the range 400 MHz$<\nu<$800 GHz. (cf. Weiss 1980). As the universe expands by a factor (a_2/a_1) from time t_1 to t_2 each photon is redshifted by the expansion factor, thus $\nu_2 = (a_1/a_2)\nu_1$. The blackbody radiation spectrum retains the form of eq. (4) above with $T_2 = (a_1/a_2)T_1$. Let a civilization measure the cosmic microwave background and determine its temperature, T. Then let that civilization emit a signal with frequency ν_e given by $h\nu_e = kT_e$. When the photon is received by an observer its frequency is measured to be $\nu_r = (a_e/a_r)\nu_e$ and at that epoch the observer will see a blackbody radiation spectrum with $T_r = (a_e/a_r)T_e$. Thus the observer finds $h\nu_r = kT_r$. Once a photon is emitted with $h\nu = kT$ it will retain that value always as the universe expands, since the other photons in the blackbody radiation redshift in exactly the same fashion. Civilizations can send signals with $h\nu_0 = kT$ knowing that they can be received at any time in the future, potentially reaching a great number of galaxies. Observers can search the entire sky at $h\nu_0 = kT$ and detect signals from galaxies regardless of redshift. With $T = 2.7$ K, $\nu_0 = 56$ GHz. The earth has a velocity of approximately 300 km s^{-1} with respect to the rest frame established by the cosmic blackbody radiation. This produces a fractional variation $\delta T/T \sim V_p/c \sim 10^{-3}$ as a function of direction. However once the local motion has been determined from the blackbody data it can be taken out and a rest standard relative to the microwave background established. If one wishes to observe in a direction X one uses $\nu_0 = kT(X)/h$ where T(X) is the value of the background temperature in direction X. For sending signals one uses $\nu_0 = kT(-X)/h$ where -X is the direction antipodal to X. This technique should reduce the uncertainty in ν_0 to one part in 10^4, the expected level of cosmological fluctuations on small and intermediate angular scales. (Gott and Rees 1975, Zeldovich 1972). Fluctuations of similar magnitude can be produced by scattering of the microwave photons in gas in clusters of galaxies. Thus an advanced civilization should be able to determine ν_0

to an accuracy of one part in 10^4 in any direction. The proposed Cosmic Background Explorer satellite (expected launch date 1987) should measure the microwave background to sufficient accuracy to allow us to establish ν_0 to one part in 10^3, (J.C. Mather, Private Communication). This requires us to search a bandwidth of$\pm \delta \nu \sim 10^{-3} \nu_0 \sim \pm 50$ MHz. With only our present knowledge (T = 2.7°K $\pm 0.25^\circ$K) we would have to search a band of ± 5 GHz about 56 GHz.

The simplest cosmic frequency standard would be $h\nu_0 = kT$. In these units the blackbody spectrum has a particularly simple form: at low frequencies $I(\nu)d\nu \propto \nu 2$(Rayleigh-Jeans Law), ν_0 is the frequency at which $I(\nu)d\nu$ falls a factor of $(e-1)$ below the Rayleigh-Jeans Law. We expect civilizations to know of the pure number e (the base of the natural logarithms) and for kT to be a natural unit of energy. A gas in thermal equilibrium with the black body radiation will have a mean energy per particle of 1/2 kT per degree of freedom. A monatomic gas will have a mean energy per particle of 3/2 kT. Thus the frequencies ν_1 = 28 GHz=(kT/h)/2, ν_0=56 GHz=kT/h, ν_2=84 GHz=3(kT/h)/2, naturally suggest themselves. The maximum intensity of the microwave background in frequency units $I(\nu)_{max}$ occurs at $h\nu_{max}$=2.81214 kT (158 GHz). The maximum emission in wavelength units occurs at $h\nu_3 = 4.9651$ kT (278 GHz). The maximum photon emission in wavelength units occurs at $h\nu_4 = 3.9207$ kT (200 GHz). The mean energy per photon in the blackbody spectrum is $h\nu_5 = 2.7011$ kT (151 GHz). (The frequency values above assume T=2.7°K, since T is currently known to an accuracy of about 10 percent the frequencies are also currently known to the same precision. Ultimately the frequency standards could be established to an accuracy of one part in 10^4). With suitable imagination perhaps a dozen plausible transmission frequencies associated with the microwave background could be found.

The minimum detectable flux P_r (Wm^{-2}) with a required signal to noise ratio of S/N is given by

$$P_r = \frac{kT_s}{A_e} \frac{\Delta \nu \, (S/N)}{\sqrt{\Delta \nu \tau}} \tag{5}$$

where A_e ($=\eta \pi r^2$) is the effective area of the antenna, $\Delta \nu$ is the bandwidth of a channel and τ is the detection time. $\eta = 0.9$ is the efficiency of the antenna and r is the radius. T_s is the system noise temperature, the sum of the background noise temperature T_N plus that due to amplifier noise, antenna imperfections and atmospheric noise. We assume that the signal has an intrinsic bandwidth that is less than our channel bandwidth. In space, for $\nu > 3$ GHz

$$kT_N \simeq \frac{h\nu}{\exp(h\nu/kT)-1} + h\nu \tag{6}$$

The first term is due to the microwave background radiation and the second is due to quantum noise. (cf. Oliver 1977). For $h\nu < kT$, $T_N \sim T = 2.7^\circ$K, for $h\nu > kT$, $T_N \sim h\nu/k > 2.7^\circ$K. Thus at frequencies significantly higher than $\nu_0 = 56$ GHz, detection becomes appreciably more difficult due to the quantum noise: $T_N(\nu_1) = 3.4^\circ$K, $T_N(\nu_0) = 4.3^\circ$K, $T_N(\nu_{max})$ =8.1°K.

This makes the frequencies ν_1 = 28 GHz and ν_0 = 56 GHz preferable to $\nu_{max}, \nu_2, \nu_3, \nu_4, \nu_5$.

Because of a wide O_2 absorption band at 60 GHz a search at ν_0= 56 GHz must be carried out from space, while 28 GHz is observable from the ground. Table 14.2 lists parameters for 4 possible experiments of varying expense. Experiments II and IV could be done with our current knowledge of the microwave background temperature. We must search a band of ± 5 GHz about 56 GHz or a band of ± 2.5 GHz about 28 GHz. After the COBE satellite has measured the microwave background temperature to an accuracy of one part in 10^3 we need only search a frequency band of ± 50 MHz about ~ 56 GHz. Experiments I and II are identical except experiment I using the COBE results takes 10^{-2} of the time to complete an all sky search because it must search a frequency band only 10^{-2} as large. Experiment III is a larger space experiment which also utilizes the COBE results. Experiment IV could be done from the ground. The flux limits reached by each experiment are given in Table 14.2. For each of the flux limits in Table 14.2 we can compute the minimum required transmitter powers for detection at the median civilization distance of 1.2×10^9 light years (as calculated in Table 14.1). Experiments I and II could detect total omnidirectional transmitter powers of 4×10^{30} W at a distance of 1.2×10^9 light years and experiments III and IV could detect total omnidirectional transmitter powers of 6×10^{28} W and 10^{33} W respectively at the same distance. Included in these calculations is the $(1 + z)^4$ decrease in brightness of the transmitters due to the redshift effect. The redshift z is given to good approximation by the formula $z = H_0 d/c = d/6000$ Mpc for z < 0.2.

Table 14.2. Possible Experiments

	I	II	III	IV
Frequency	~ 56 GHz	~ 56 GHz	~ 56 GHz	~28 GHz
Searched	± 50 MHz	± 5 GHz	± 50 MHz	± 2.5 GHz
Dish Diameter		2.4 m	20 m	0.3 m
Location		Space	Space	Ground
Bandwidth		5Hz	5 Hz	2.5 Hz
No. of Channels		10^6	10^6	10^6
τ		0.2 s	0.2 s	0.4 s
T_{sys}		10°K	10°K	100°K
All Sky Search	83 days	23 yrs.	14 yrs.	64 days
Minimum Detectable Flux with S/N = 10 in W m^{-2}		2×10^{-21}	3×10^{-23}	5×10^{-19}

What transmitter powers are possible? Dyson (1960) has described how an advanced civilization can utilize the entire energy of its sun by surrounding it with solar power stations (a Dyson sphere). A mega-civilization which has colonized a galaxy could tap the energy of all of its stars using Dyson spheres. Kardashev (1964) classifies civilizations by their power output: a type I civilization $= 4 \times 10^{12}$ W is like the earth, a type II civilization $= 4 \times 10^{26}$ W $= 1$ $L_\odot$ utilizes the energy of a star like the sun, a type III civilization $= 4 \times 10^{37}$ W $= 10^{11} L_\odot$ utilizes the energy output of an entire galaxy. In the colonizing models we have discussed in section III the median mega-civilization we see contains 5×10^{10} inhabited planets. If each of the 5×10^{10} colonies utilized 1 percent of the total energy of its own sun and used 1 percent of this energy for transmitting signals then the mega-civilization would produce a total transmitter power of $10^{-4} \times 5 \times 10^{10} L_\odot = 2 \times 10^{33}$ W which could be detected by all of the experiments. (Perhaps the most efficient way for such a mega-civilization to produce a large isotropic transmitter power would be to build a large number of highly directional antennas. Highly directional beams could be sent out by each colony within the mega-civilization for example). If all civilizations devote the same fraction of their energy resources to transmitting, a type III civilization is as detectable at a distance of 1.2×10^9 light years as a type II civilization is at a distance of 4000 light years and as a type I civilization is at a distance of 4×10^{-4} light years. Thus even if there are other civilizations in our galaxy the most easily detected one may be extra-galactic.

To summarize, the proposed experiments in Table 14.2 could detect total transmitter powers in the range 6×10^{28} W to 10^{33} W at a distance of 1.2×10^9 light years. If colonization is allowed we expect the median mega-civilization to contain 5×10^{10} inhabited planets. Such a mega-civilization could produce a total transmitter power of 2×10^{33} W by utilizing 10^{-4} of its potentially available energy resources. This would be detectable by all four experiments. Thus searches for extra-galactic civilizations at 28 and 56 GHz can be a useful and important compliment to traditional searches for other civilizations within our galaxy.

V. CONCLUSIONS

In this paper, detectability of civilizations at cosmological distances is examined. The first step is to calculate the number of habitable planets. If there are of order 10^9 habitable planets in our galaxy, the mean cosmological density of habitable planets is of order 3×10^6 HP Mpc^{-3} (with $H_0 = 50$ km s^{-1} Mpc^{-1}). If the universe is simply connected, then the number of habitable planets in the universe is $> 10^{19}$ if $1 < \Omega < 2$ and is infinite if $\Omega \leq 1$. Let P_i be the probability of an intelligent civilization arising spontaneously on a habitable planet. We are not unique if P_i is greater than 10^{-19}. If civilizations began forming 2 billion years ago we have the possibility of seeing other civilizations if

$P_i > 1.5 \times 10^{-15}$. The number of visible civilizations as a function of P_i is calculated in Table 14.1. If colonization is normal the number of visible civilizations can be as high as 4×10^{13} even if P_i is relatively low. This upper limit is independent of the speed of colonization. The fact that the earth has not been colonized only means that the spheres of colonization have not overlapped by the present epoch; in the most favorable scenarios up to half the habitable planets could have been colonized. This is a conservative upper limit because it assumes colonies will only be founded on habitable planets, whereas with the use of space colonies a wealth of additional sites may be utilized. Thus even if we are alone in our galaxy there may be many civilizations visible in external galaxies. If $P_i > 10^{-9}$ other civilizations form in our galaxy. In this case colonization is not allowed, and the number of extragalactic civilizations visible is 7×10^5 times larger than the number of civilizations in our galaxy. So if intelligent life forms frequently there are still many more opportunities to detect life at large distances. The results are summarized in Table 14.1.

For communication at cosmological distances line radiation such as 21-cm is not very useful because of the cosmological redshift. However, we can establish frequency standards tied to the $T = 2.7^\circ K$ cosmic microwave background discovered by Penzias and Wilson (1965). Two promising frequencies are 28 GHz ($h\nu = 1/2$ kT) and 56 GHz ($h\nu = kT$). With careful measurements of the microwave background an advanced civilization could set the frequency standard to an accuracy of one part in 10^4. Several experiments of varying expense are outlined in Table 14.2. A space experiment is required to observe 56 GHz but modest searches at 28 GHz can be conducted from the ground with present technology.

For purposes of illustration we have adopted specific values for various parameters and tabulated the results as a function of P_i. From the information given it should be clear how to construct similar tables for a variety of adopted parameter values. Even if there are other civilizations within our galaxy, the one most easily detected may be extragalactic. If the probability of intelligent life developing spontaneously is small then extragalactic searches are the only ones with a chance of success. Even if we are the only civilization in our galaxy, there are plausible colonization models in which there are many civilizations visible in external galaxies. The points mentioned above show that it is useful to find strategies to search for life at cosmological distances. We propose frequency standards tied to the cosmic microwave background radiation. Two plausible ones are 28 GHz and 56 GHz. Modest searches at 28 GHz could be done from the ground with present technology.

REFERENCES

Bracewell, R.N. (1975). The Galactic Club: Intelligent Life in Outer Space. W.H. Freeman, San Francisco, Distributed by Scribner, New York.

Cocconi, G. and Morrison, P. (1959). Nature, 184, 844.

Dyson, F.J. (1960). Science, 131, 1667.

Dyson, F.J. (1979). Reviews of Mod. Physics, 51, 447.

Gott, J.R. and Rees, M.J. (1975). Astron. & Astrophys., 45, 365.

Gott, J.R. and Turner, E.L. (1976). Astrophysical Journal, 209, 1.

Gott, J.R. (1980). Monthly Notices Royal Astronomical Soc., 193, 153.

Hart, M. (1975). Quarterly Journal Royal Astronomical Society, 16, 128.

Kardashev, N.S. (1964). Soviet Astronomy-AJ, 8, 217.

Mather, J.C. (1980). Private communication.

Oliver, B.M. (1977). In The Search for Extraterrestrial Intelligence, NASA SP-419, 63.

Peebles, P.J.E. (1974). Astrophysical Journal, 189, L51.

Penzias, A.A. and Wilson, R.W. (1965). Astrophysical Journal, 142, 419.

Shklovskii, I.S. and Sagan, C. (1966). Intelligent Life in the Universe. Holden Day, San Francisco.

Sandage, A. and Tammann, G. (1975). Astrophysical Journal, 197, 265.

Weinberg, S. (1972). Gravitation and Cosmology. Wiley, New York.

Weiss, R. (1980). Ann. Rev. Astron. & Astrophysics, 18, 489.

Zeldovich, Ya. B. (1972). Monthly Notices Royal Astronomical Soc., 160, 1P.

15

Nucleosynthesis and Galactic Evolution: Implications for the Origin of Life*

Virginia Trimble

ABSTRACT

The terrestrial example suggests that the evolution of intelligent life requires large amounts of both heavy elements (those beyond hydrogen and helium) and time. How much of each is available in various places depends on the chemical and dynamical evolution of our galaxy. Particular questions that we are beginning to be able to answer and that probably have some bearing on "Where Are They?" include (a) how old is the universe as a whole? (b) what is the age of the oldest stars in our Galaxy? (c) what is the maximum heavy element abundance found among the globular cluster stars? (d) what is the distribution of ages of disc stars? and (e) what is the distribution of heavy element abundances in disc stars, and how is it correlated with ages? Though details of each of these can be debated, the most probable conclusion is that stars with both ages and heavy element abundances comparable with those of the solar system are quite common in the Galaxy, particularly in its inner regions. Non-detection of extraterrestrials is not explicable in terms of stellar and galactic evolution.

We cannot expect living creatures to be common in the Galaxy if planets to support them or stars to warm them are rare. Other authors in these proceedings address portions of this problem: what sorts of

---

*This talk was originally scheduled to be given by Beatrice Tinsley. I am grateful to her and to her Yale colleagues, Bruce Twarog, Pierre Demarque, and Richard Larson, for enlightening discussions and for data in advance of publication. The Institute of Astronomy (Cambridge) provided a quiet corner in which the paper was written, for which hospitality I am, as always, indebted to the Director.

stars, for instance, can provide a constant-temperature environment for billions of years, and are such stars likely to have planets within their life-support zones?

Our current understanding of the formation and evolution of the Galaxy as a whole leads to additional questions bearing closely on "Where Are They?" The general scheme is that our own and other galaxies began condensing 15-20 billion years ago out of lumps in a previously (almost) homogeneous, expanding, cooling gas of hydrogen and helium (the accepted products of the standard Hot Big Bang model of the very early stages of the universe). Within those lumps, smaller lumps gradually turned into stars. And in the stars (particularly those much more massive than our sun) there occurred nuclear reactions that turned some of the hydrogen and helium into all the other elements, after which, assorted explosions dispersed the newly-created heavy elements back into the remaining gas, from which later generations of stars could form.

Astronomers call everything except hydrogen and helium "metals" for historical reasons. Our excuse for foisting this terminology on the rest of the world is that one gets awfully tired of saying "heavy elements" and "elements beyond hydrogen and helium." The "metal" abundance by mass is usually called Z; and our sun has $Z_o \sim 0.02$. Abundances relative to the sun can be expressed as Z/Z_o.

Now, among the elements born in stars are ones essential to chemically-based life: the carbon, oxygen, nitrogen, calcium, iron, gold, and all the rest, without which we could not exist or build our civilizations. Thus, surely, the first generation of (pure hydrogen and helium) stars could not have had life-bearing planets. Only after nuclear reactions had built the metal supply up to some minimum level could earth-like planets form. And only some long time after that can we expect to find advanced life forms. We do not, unfortunately, know how much carbon (and so forth) is enough nor how much time is required. In what follows, it will be assumed that the requirements are metal abundances within a factor of two or three of solar and times comparable with the age of the earth.

Thus, we would like to know (a) what percentage of the 2×10^{11} galactic stars are more than 4.65 billion years old, (b) how many of them contain at least 1 percent metals, and (c) if there are such stars, where are they? Almost any astronomer asked these questions in recent years would have known an answer. Unfortunately, different astronomers have known different answers at different times. What follows is, therefore, an attempt to outline our current understanding of galactic chemical evolution and how it has developed.

Within our Milky Way Galaxy and others of similar, spiral, type, the condensation and fragmentation processes described above must have occurred in at least two phases. (More complex scenarios are possible, but do not change the answers to questions a, b, and c.) First in our lump (or protogalaxy) came an approximately spherical distribution of stars called the galactic halo. Many halo stars (and the ones easiest to study) are in compact groups called globular clusters, with 10^5-10^6

stars per cluster. The oldest of the 100+ globular clusters in our galaxy probably reach or exceed 15 billion years in age. None is likely to be as young as 5 billion years. Most halo stars are deficient in metals, relative to the sun, by factors of 10-100, and so probably cannot have terrestrial planets. The maximum heavy element abundance in these stars is rather difficult to determine partly because they are very far away and partly because their great age means that they cannot easily be calibrated against nearby, better-understood stars. The maximum heavy element abundance in globular cluster stars is one of the issues on which the astronomical community is not entirely in agreement.

Some time after the last halo stars formed (how long after is another of those disputed issues) remaining gas, now somewhat enriched in metals (just how much is also uncertain) flowed down into a disc at the center of the halo and began fragmenting into additional stars. Star formation in the disc continues to the present. The things we need to know about disc stars are: when did they begin to form; how rich in metals were the first ones; and how have the average and maximum metal abundances of the stars being formed changed since then?

Figure 15.1 shows the standard picture of galactic chemical evolution as presented in most textbooks and review articles until quite recently (see, e.g., Trimble 1975 for further details). In it, the globular clusters are all formed at about the same time, with metal abundances, Z, covering the range ~ 0.001 to ~ 0.5 times solar. Disc stars started forming immediately thereafter, with metal abundances about equal to the maximum of those in the globular clusters. Star formation then continued at roughly constant rate and with the average Z increasing linearly with time, though with factor of two variations about the average at any given instant (depending primarily on location in the galactic plane). The sun, being average for its age, will, according to this model, be matched in metal abundance by a large number of stars of the same or slightly larger ages. Something like this model is implicit in the numbers usually fed into the Drake equation.

Two changes in this picture have been suggested, which, together, would considerably reduce (perhaps to zero) the number of stars that could have planets with advanced life forms at the present time. First, Cohen (1980) analyzed the spectra of giant stars in several of the highest-Z globular clusters and concluded that their metal abundances were considerably lower than previously thought. Second, the Yale group (see Tinsley and Larson 1977) suggested (from detailed comparisons of star colors and brightnesses with calculated evolutionary tracks) that the oldest disc stars were not much older than the solar system. This implies a long hiatus between the cessation of halo star formation and the beginning of disc star formation, although the gas metallicity continued to increase during the latency period. Such a model, as shown in Figure 15.2, could explain a nearly total absence of complex living creatures: the halo stars having too few metals and the disc stars too little time for their development. Thus, life might, in principle, be quite probable in the Galaxy, but we would have to come back in a few billion years to observe its effects.

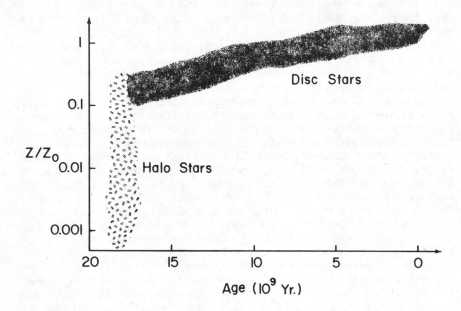

Fig. 15.1. Canonical model of galactic chemical evolution (for further details see Trimble 1975). In this model, globular clusters are all about the same age, but have a wide range of heavy element abundances, Z, ranging almost up to solar (Z/Z_0 = 1). Disc stars have been formed approximately uniformly through time (for more than twice the age of the Solar System), with gradually increasing mean metallicity. Under these circumstances, stars both old enough and rich enough in metals to have advanced life forms could be common.

That was the situation at the time this meeting was organized; it was evidently hoped that considerations of galactic evolution might provide an out for those who regard absence of evidence (for extra-terrestrials) as evidence of absence, but who would, nevertheless, like to think that we are not really special.

But even this small comfort was to be short lived. Further work on stars in high-Z globular clusters (Manduca 1980) suggests that, although they are not quite as metal rich as once advertised, the change is a rather small one. In addition, Demarque (1980) finds that abundances in clusters are a function of age, the youngest, highest-Z globulars joining smoothly onto the oldest, lowest-Z disc stars. Finally, the oldest disc stars (Twarog 1980) are once again thought to go back more than 10 billion years, with a small but systematic increase in metallicity as a function of time of formation.

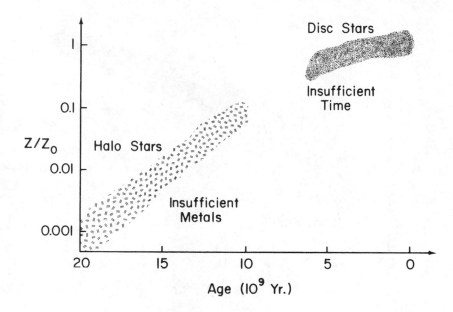

Fig. 15.2. Revised model of galactic chemical evolution, incorporating the low maximum metallicity found by Cohen (1980) for globular cluster stars and the apparent absence of old disc stars suggested by the Yale isochrones (Tinsley and Larson 1977). Such a galaxy would have few or no stars both old enough and rich enough in metals to support intelligent life on their planets.

Thus, we are back nearly where we started from with the canonical model: stars with $Z/Z_0 \gtrsim 0.5$ and ages ≥ 4.65 billion years should be quite common, although no clusters of disc stars much older than the solar system have been found. Figure 15.3 shows the current (1980.7) situation (see Twarog 1980 for further details on numbers of disc stars as a function of age and the like).

Finally, we might ask where to look for high-Z oldish stars. Both average metal abundance and average age depend somewhat on position in the galaxy, particularly distance from the center. Figure 15.4 is a somewhat impressionistic view of the relationships (which have not been evolving as rapidly with time as some of the other issues here discussed). The data are given more precisely by Trimble (1975), Tinsley (1980) and Pagel (1981). In general, the most promising stars will be located closer to the galactic center than we are.

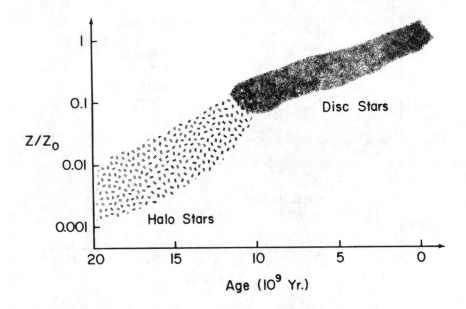

Fig. 15.3. The 1980.7 model of galactic chemical evolution. Although some details are different from the canonical version (Fig. 15.1), once again there seem to be plenty of stars of solar or higher age and approximately solar Z. For further details see Tinsley (1980) and Pagel (1981).

REFERENCES

Cohen, J.G. (1980). In Star Clusters, Proc. of IAU Symp. No. 85, (editor: J. Hesser). D. Reidel, Dordrecht, Holland, 385.

Demarque, P. (1980). In Star Clusters (editor: J. Hesser), op. cit., 218.

Manduca, A. (1980). Ph.D. dissertation, University of Maryland.

Pagel, B. (1981). Proceedings of NATO Advanced Study Institute, The Structure and Evolution of Normal Galaxies. Cambridge University Press.

Tinsley, B.M. (1980). Fund. Cos. Phys. 5, 287.

Tinsley, B.M. and Larson, R. (editors), (1977). The Evolution of Galaxies and Stellar Populations. Yale University Observatory, New Haven.

Trimble, V. (1975). Rev. Mod. Phys. 47, 877.

Twarog, B. (1980). Ph.D. dissertation, Yale University.

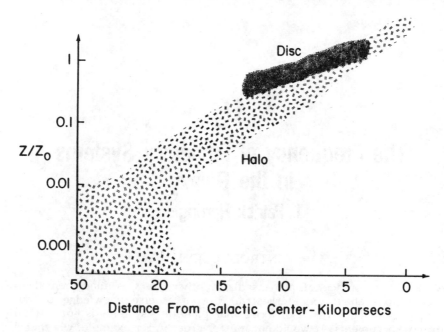

Fig. 15.4. Gradients of average metal abundance as a function of distance from the galactic center in halo (plus bulge) and disc populations of the Milky Way.

DISCUSSION

Robert Rood: I think that we should be especially careful in comparing stellar ages determined by different methods. The Sun is different from all the other stars and, working in this field, I would say that probably there is a 50 percent uncertainty generally in attaching other stellar ages to the solar age. So I would put 50 percent error bars on all those ages, in which case you get no information whatsoever.

Trimble: That isn't quite true.

Rood: Working with solar neutrinos gives me a great deal of skepticism in this area.

16

The Frequency of Planetary Systems in the Galaxy

J. Patrick Harrington

INTRODUCTION

While the existence of planets about other stars is fundamental to speculations about extraterrestrial life, a precise knowledge of the frequency of occurrence of extrasolar planetary systems is not crucial to most arguments. It would not matter greatly, for example, whether 5 percent rather than 50 percent of all stars were accompanied by planets. It would matter, however, if only one star in 10^6 has a planetary system. So the key question is whether or not planetary systems are rare.

One may feel reluctant to entertain the hypothesis that the solar system is a highly unusual phenomenon. Nevertheless, this feeling cannot be justified simply by appealing to the "Copernican principle" that we do not occupy a privileged position in the universe. Since the existence of the solar system is a precondition of our own existence, we can hardly regard our observation of it as a statistically random sample of the universe at large. We must either find empirical evidence for the existence of extrasolar planets, or we must be able to show how the formation of a planetary system fits into the normal process of star formation. We will look at these two options in turn.

OBSERVATIONAL EVIDENCE FOR EXTRASOLAR PLANETS

In the visual region of the spectrum, planets are fainter than the stars they orbit by factors of $\gtrsim 10^9$, so we cannot hope to observe them directly by conventional means. The most promising indirect method for the detection of planets is the search for small irregularities in the motion of the star due to the gravitational effects of its invisible companion. This "astrometric method" should eventually provide definite results (Gatewood, 1976; Gatewood et. al., 1980).

From time to time, various investigators have in fact reported small deviations from rectilinear motion, and have interpreted this as evidence of unseen companions. Most of these cases have subsequently been disallowed. The best case has usually been thought to be that of Barnard's star, for which companions comparable in mass to Jupiter have been claimed (van de Kamp, 1975). It is now accepted that the original 25 year period was partly an artifact of optical changes in the telescope. After allowance for these changes, small perturbations remain. The nature of these perturbations is controversial, but there are indications that these remaining effects may be due to systematic errors. Heintz (1978) recently examined the available astrometric evidence for the existence of stars with low mass companions and concluded that "There is now no observational evidence for any companion less massive than the lowest-mass <u>visible</u> stars known (0.06 solar masses in Wolf 424)."

Negative astrometric results are not without significance, however, for they demonstrate that it is not unusual to find stars that are effectively single (i.e., stars which have no companions with masses much above the known range of planetary masses). Barnard's star, for example, could not have a 0.01 solar mass companion anywhere nearby. This result is not trivial because of the large but uncertain fraction of stars found in binary systems.

Other methods for the detection of extrasolar planets have been proposed. One may look for the very small (± 10 m/s) variations in the velocity of the star as it orbits the center of gravity of the star-planet system (Serkowski, 1976). This will require an order of magnitude improvement in the accuracy of such measurements. Another approach would take advantage of the increase in the (planet/star) brightness ratio in the infrared as compared to visual wavelengths (e.g., Bracewell and MacPhie, 1979). Observations from above the earth's atmosphere offer further possibilities for the future. The first such opportunity will be provided by the Space Telescope. While this instrument will <u>not</u> be able to see extrasolar planets directly, it should be able to measure stellar positions more accurately than ground-based instruments and thus could search hundreds of stars for planetary companions by the astrometric method (Baum, 1980). These ideas have yet to be exploited. However, when taken together with improvements in ground-based astrometry, an observational answer to the question of the frequency of planetary systems does not seem too far away. But in the meantime, we must turn to arguments based upon our ideas of the origin of the solar system.

INDIRECT ARGUMENTS FOR EXTRASOLAR PLANETS

The Origin of the Solar System

It is widely accepted that the solar planetary system grew within a rotating, flattened nebula of gas and dust and that this nebula was coeval with the formation of the sun itself, which at some point developed as the central condensation of this nebula. Such a picture is strongly supported by the regular structure of the solar system and by the coincidence in the ages of the oldest planetary material and meteoritic debris. Consensus disappears, however, when we try to fill in the details of this picture. A good account of current theories in all their complexity can be found in the recent volume Protostars and Planets (edited by Gehrels, 1978; hereinafter "P&P").

Two possible routes to the formation of planets have received attention. Along the first route, the collapse of an interstellar cloud results in a massive (more than one solar mass) nebula which at first has little in the way of a central condensation, because the angular momentum of the material prevents collapse towards the axis of rotation. Before the transfer of angular momentum within this pre-solar nebula can allow much material to move to the center and complete the formation of the sun, instabilities in the disc lead to the formation of large, self-gravitating protoplanets. The terrestrial planets may form at the core of the inner protoplanets, which eventually lose their gaseous mantle to the growing sun, while the outer protoplanets develop directly into the gas giants of the outer solar system. This is the route which is now being explored by Cameron (P&P, 453).

The second route begins with the substantially complete central condensation surrounded by a much less massive ($\lesssim 0.1$ solar masses) nebula. Under these conditions the nebula will be stable and self-gravitating protoplanets cannot form. Instead, the dust component may settle to the midplane of the nebula and this thin sheet of solids can clump under self-gravitation. Or perhaps purely mechanical sticking of particles as they collide (like wet snowflakes) will result in the growth of larger objects. In any case the result is myriads of small solid planetesimals near the nebular midplane, which slowly aggregate into bigger objects until gravitation accelerates the final sweep-up by the largest body in each zone. This second route is inherently more time consuming than the protoplanet route.

Most astronomers favor the second route. The physical characteristics of meteoritic material in particular is more easily explained by growth through a planetesimal stage. Data increasingly indicate that a substantial amount of dust has actually survived from the interstellar medium and has been incorporated into meteorites without vaporization and recondensation (Clayton, 1980).

A fundamental difference between the two models is their amount of angular momentum. Cameron's massive nebula has an angular momentum of $J=8 \times 10^{53}$g cm^2 sec^{-1}. This is about the amount of angular momentum that would be expected to result from the collapse of a solar mass of material from the interstellar medium if we do not invoke any special process to remove angular momentum during the collapse. The second class of models in which planetesimals form have much less angular momentum. The angular momentum of the solar system today is $J=3 \times 10^{50}$ g cm^{2}sec^{-1}, which is almost entirely due to the four giant outer planets. If, however, we increase the masses of the planets to allow for the lost volatiles according to the chemical composition of the sun we find that the early solar system must have had at least $J=3\times10^{51}$ g cm^{2}sec^{-1}. The second class of models generally have angular momenta in the range 3×10^{51} to 5×10^{52}. The pathway, therefore, from interstellar clouds to planets is not a smooth one; there is an angular momentum discontinuity which must somehow be bridged.

Stellar Rotation and Planetary Systems

Solar-type stars, and lower main sequence stars in general, are slow rotators. The sun itself turns with an equatorial velocity of only 2 km sec^{-1}. This is in contrast to the more massive stars of the upper main sequence which rotate much more rapidly with velocities >100 km sec^{-1}. When the rotational velocities of stars on the main sequence are plotted as a function of spectral type, an abrupt drop in velocity is seen to occur near spectral class F5. It was noted, however, that while the sun itself has much less angular momentum than the upper main sequence stars, the angular momentum of the solar system as a whole is comparable to that of those stars. Thus the hypothesis was advanced that stars later than F5 are accompanied by planetary systems, which when they form take up the angular momentum which would otherwise appear as stellar rotation (Huang; 1957, 1965). The implication was that planetary systems are universal for stars below spectral type F5.

This argument lost its force, however, with the discovery that the rotational velocity of the lower main sequence stars depends strongly on the age of the star. In fact young stars, like the T Tauri stars and stars in the young cluster NGC 2264, have rotational velocities of 50-150 km sec^{-1}. It appears that the spin-down of stars continues throughout their lifetime, closely following a $t^{-0.5}$ law over ages ranging from 10^6 to 10^{10} years (Skumanich, 1972; Imhoff, P&P, 699). This gradual braking of rotation is doubtless due to mass loss by stellar winds which are coupled to the rotating star by magnetic fields. Thus the loss of angular momentum occurs gradually <u>after</u> the supposed planets would have formed, so that the slow rotation of the old lower main sequence stars cannot establish the existence of planetary systems.

Infrared Observations of Young Stars

Infrared observations of young stars and protostars are important because they can provide information about dust in the vicinity of stellar or pre-stellar objects. Of particular interest are T Tauri stars, which for some time have been known to radiate more strongly in the infrared than normal stars of their spectral class. While some of this radiation may arise in an extended chromosphere near the surface of the star, silicate emission features at 10 μm have been observed which confirm the presence of circumstellar dust. This dust would appear to have a temperature around 200K and would be located a few astronomical units from the star (Rydgren, Strom and Strom, 1976; Rydgren, P&P, 690). Polarization has been observed which shows that the distribution is not spherically symmetric (Bastien and Landstreet, 1979); this would be consistent with our expectation that the circumstellar material would relax to a disc. The interpretation of the infrared data is as yet beset with uncertainties, but it at least seems clear that the dust laden clouds needed for the formation of planets are sometimes present. However, as Hartman (P&P, 58) has emphasized, while it is encouraging to have observational evidence of dust around the youngest stars, we simply do not know whether the eventual clearing of this dust signifies the accretion of small particles into planetesimals or merely the dispersal of circumstellar material by a stellar wind.

Binary Stars and Planetary Systems

While we do not know how frequently stars are accompanied by planets, we do know that in most cases they are accompanied by other stars. As far back as the pioneering work of Kuiper (1935), one of the motives for gathering data on multiple stars has been a desire to shed some light on their relation to possible planetary systems. The recent work of Abt and Levy (1976; also the revisions in Abt, P&P, 323) is noteworthy. They studied a rather complete sample of solar-type stars (123 main sequence stars of spectral type F3-G2, after excluding 12 high velocity stars) and made careful estimates of incompleteness effects. They conclude that 54 percent of all solar-type stars have close stellar companions (periods less than 100 years) and that, in addition, 53 percent have distant stellar companions (periods greater than 100 years). (The sum exceeds 100 percent because there are many triple systems.)

    The distinction between close and distant companions stems from their observation (see also Trimble & Cheung, 1976) that the distribution in mass for the secondary star behaves differently for these two groups: The close binaries are most likely to have a secondary of the same mass as the primary, with the number of pairs, N, with secondaries of mass m_2 declining according to $dN \propto m_2^{0.4} \, d(\ln m_2)$. The distant binaries, on the other hand, are likeliest to have a secondary of much smaller mass; the mass distribution of the secondaries is indistinguishable from that followed by the primaries and by other field stars in general – the van Rhijn distribution. Abt and Levy make the reasonable proposal that the distant binaries formed to all intents and purposes as isolated stars, but were near enough to remain gravitationally bound, while the close binaries are the result of the bifurcation of what started as a single protostar. They also find that the number of binaries as a function of period has a single broad maximum, with a median period of 14 years. Thus the median separation of a binary is about the same as the distance between the sun and the giant planets in the solar system.

    What can this tell us about the possibility of extrasolar planets – other than that stars frequently have another star where the planets ought to be? One approach, adopted by Abt and Levy, is just to regard stars with planetary companions as binaries with very low mass secondaries. They thus extrapolate their $m_2{}^{0.4}$ relationship to $m_2 = 0$ with the following results: 54 percent of solar-type stars will have close stellar companions, 13 percent will have "black dwarf" companions, 11 percent will have planetary companions, and 22 percent will be left without companions of any sort. (A "black dwarf" is here defined as an object in the 0.07 to 0.01 solar mass range. There is no real evidence for or against the existence of such objects – see the Heintz quote above.)

While this approach has the advantage of giving us a numerical result, not everyone will accept the implied lack of distinction between binary stars and planetary systems (e.g., Huang, 1977). Extrapolating the uncertain $m_2{}^{0.4}$ relationship from the smallest observed stellar companions of about 0.1 solar mass down to Jupiter's mass – a factor of 100 – seems bold. Furthermore, the origin of binary stars is probably qualitatively different from that of planets. We believe the planets grew slowly by accretion in a disc stabilized by a pronounced central condensation. Binary stars, on the other hand, seem to have formed much more rapidly, on a time scale comparable to their orbital period. The eccentric nature of their orbits is evidence of this. Fig. 16.1 shows the binaries from Abt and Levy's study with periods between 20 days and 100 years, plotted against their orbital eccentricity. The mean eccentricity is nearly one half. One possibility is that the infall of the material had not been stopped by angular momentum induced rotation at the time the two stars established their identity. In any case the process would be quite different from the slow formation of a system of planets in nested, coplanar, circular orbits.

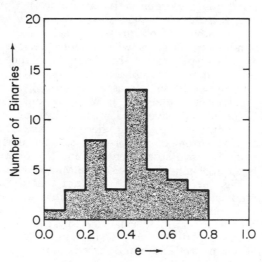

Fig. 16.1. The eccentricities of binaries with periods between 20 days and 100 years (Abt and Levy, 1976). The mean eccentricity of the sample is 0.42.

The numerical studies of protostellar collapse now being actively pursued are illuminating. Because the 3-dimensional treatment of hydrodynamical collapse presents severe numerical difficulties, most calculations which include the effects of rotation have imposed the constraint of axial symmetry. These calculations usually show that the angular momentum keeps the material away from the rotation axis, so that the density maximum does not occur at the center but instead a ring-like structure forms (Bodenheimer and Black, P&P, 228; but also see Norman, Wilson and Barton, 1980). It is generally believed that without the constraint of axial symmetry such rings would fragment and probably form a binary system. This is just as well, for a theory of star formation should explain why most stars are in binary systems; such a theory need not routinely yield planets. Safronov and Ruzmaikina (P&P, 545) suggest that the outcome of cloud collapse depends upon the angular momentum of the cloud: High values may produce binaries, but if the angular momentum is less than about 10^{53} g cm^2 sec^{-1} a central condensation will form, surrounded by a disc suitable for accretion of planetesimals.

Since angular momentum seems to be the key factor in planet formation, it may be useful to look at the momentum distribution found in binary systems. The angular momentum per unit mass (specific angular momentum) is given by

$$j = 4.46 \times 10^{19} \frac{m_1 m_2}{(m_1+m_2)^{4/3}} P^{1/3} (1-e^2)^{1/2} \ cm^2 \ sec^{-1},$$

where m_1 and m_2 are the masses of the stars in solar units, P is the period in years and e is the eccentricity of the orbit. Unfortunately, for the single-lined spectroscopic binaries we only know the masses in a statistical sense. However, to obtain some idea of the j distribution, we have taken the numbers of stars with various periods from Abt and Levy's study of solar-type stars (Abt, P&P, 323) and integrated that against their derived mass distribution, $dN \propto m_2^{0.4} d (\ln m_2)$ over the range $m_2 = m_1$ to $m_2 = 0.05$ solar masses, to obtain the relative number of binaries with different values of j. We have only considered close binaries (P < 10^5 days) and have used mean values of m_1 = 1.3 and e = 0.4. The result is shown in Fig. 16.2, where we have also indicated the specific angular momentum of the present and the early solar system, of Cameron's massive pre-solar nebula, and of observed molecular clouds as discussed by Field (P&P, 243). The Maxwellian distribution has been included in Fig. 16.2 to illustrate the fact that the j-distribution is really quite broad and is not simply explainable as due to statistical fluctuations about some mean value. Furthermore, the single stars not included in this distribution must raise the low-j end of the curve. It is particularly interesting that the distribution reaches down to the specific angular momenta attributed to the early solar system; this could imply that conditions suitable for the formation of planets occur frequently.

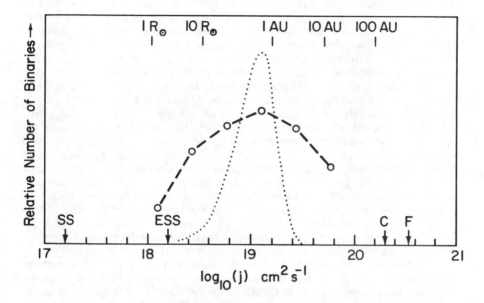

Fig. 16.2. The relative number of binaries as a function of specific angular momentum, j. (The vertical scale is linear, but otherwise arbitrary). The dotted line is a Maxwellian distribution. SS, ESS, C and F represent the present solar system, the early solar system, Cameron's massive solar nebula (P&P, 453) and observed interstellar dark clouds (Field, P&P, 243), respectively. The scale at the top is the separation of two solar-mass stars in circular orbit with the indicated j value.

We also note that the median angular momentum in a binary is more than an order of magnitude less than what one might expect in interstellar clouds. We seem to need a mechanism which will remove a large amount of angular momentum, but which can vary greatly in effectiveness. It has often been suggested that magnetic fields play this role. According to Mouschovias (P&P, 209), the field will couple the collapsing inner regions with the slowly rotating envelope until the density becomes high enough that ambipolar diffusion lets the material slip the field lines; from then on the angular momentum will be conserved. The critical density at which the separation of the magnetic field occurs will depend upon the state of ionization of the gas. It thus could depend, for example, on the local cosmic ray flux. This could explain the large spread in the j values of observed systems.

From the ideas outlined above, we might sketch the following hypothesis: Some mechanism such as magnetic braking produces collapsing cloud cores with a wide range of angular momenta. If the specific momentum is above some critical value, say $j \simeq 10^{18}$, the structure will become unstable and fragment violently to form a binary. If the specific angular momentum is below the critical value, a central condensation will grow and excess material will form a disc stable against gravitational breakup in which planets might accumulate. (Of course if there is not sufficient material in the disc, it may dissipate without the formation of any large bodies.) Since the process of planet formation probably would not use most of the material in the disc, one would not expect the distribution function for secondary masses to be continuous when passing from stars to planets. Since it appears that the smallest values of angular momenta found in binaries reaches right down to the value for the early solar system – presumably our critical value – it may be that solar-type stars never have "black dwarf" companions. It would also be expected that planetary systems would be somewhat more frequent than the 11 percent predicted by Abt (P&P, 323). The foregoing hypothesis may of course be much too simplistic; we might well expect, for example, that the distribution of angular momentum within the cloud is important, not just its total value.

So far we have implied that binaries and planets are somehow incompatible but it is not clear that this is true. There are three distinct questions: Would a given planetary orbit be stable? If it is stable, would planetary accretion occur in this orbit in the presence of the gravitational perturbations of the second star? Finally, would a disc of material even be present in a system that had fragmented to form a binary? We have a reasonably good answer only to the first question.

We know that some planetary orbits would be unstable in a binary system. But it seems that stable orbits can exist either far away from a close binary or close to either or both members of a wide pair. Harrington (1977) has made numerical studies and has found that for stability, the ratio of the periasteron distance of the outer component to the semimajor axis of the close component (regardless of which is the planet) must be larger than 3 or 4. While this does rule out some binaries of periods around a year for habitable planets, the range in binary periods is so great that most pairs are either close enough or sufficiently distant to have stable orbits in the habitable zones, so that the potential number of interesting planetary systems is not much reduced by stability considerations alone.

Even if the planetary orbit is stable, the presence of the second star can introduce periodic perturbations in the orbit. At the time when the planetesimals form, the velocities with which bodies collide is critical – if the relative velocity is too high, the objects may fragment rather than grow. Thus the perturbations introduced by a companion could prevent planet formation over a larger zone than that forbidden on stability grounds. An attempt to evaluate this effect was made by Heppenheimer (1974), who found it to be important. His results, however, are based on the nebular model of Cameron and Pine (1973), which may not be stable against gravitational fragmentation. Similar calculations could be made for other models of planetesimal formation. Hohlfeld and Terzian (1977) have extended this idea to the effects of the perturbations produced by Abt and Levy's hypothetical population of black dwarf companions.

Perhaps the most serious question is whether the fragmentation of the collapsing cloud into a binary will leave sufficient material in the vicinity of the fragments to allow viable planetesimal growth regardless of perturbations. The present state of hydrodynamical calculations is not sufficiently advanced to offer any insight into this question.

SUMMARY

Observations have not yet provided any persuasive evidence for the existence of extrasolar planetary systems. The observed angular momenta of binary stars, coupled with current ideas of how the solar system formed, suggest that most of the 40 percent or so of stars without close stellar companions may be accompanied by planets. It is doubtful whether planets could also exist in binary star systems. Such indirect arguments, however, are weak because the origin of our own planetary system is poorly understood. We still cannot dismiss the proposition that planetary systems are extremely rare.

REFERENCES

Abt, H.A. & Levy, S.G. (1976). Astrophysical Journal Supplement 30, 273.

Bastien, P. & Landstreet, J.D. (1979). Astrophysical Journal Letters 229, L137.

Baum, W.A. (1980). IAU Colloquium No. 54 "Scientific Research with the Space Telescope", (editors: M.S. Longair and J.W. Warner), NASA CP-2111, pp. 90-95.

Bracewell, R.N. & MacPhie, R.H. (1979). Icarus 38, 136.

Cameron, A.G.W. & Pine, M.R. (1973). Icarus 18, 377.

Clayton, D.D. (1980). Astrophysical Journal Letters 239, L37.

Gatewood, G. (1976). Icarus 27, 1.

Gatewood, G., Breakiron, L., Goebel, R., Kipp, S., Russel, J., & Stein, J. (1980). Icarus 41, 205.

Gehrels, T. (editor), (1978). Protostars & Planets. Univ. of Arizona Press, Tucson, Arizona.(P&P).

Harrington, R.S. (1977). Astronomical Journal 82, 753.

Heintz, W.D. (1978). Astrophysical Journal 220, 931.

Heppenheimer, T.A. (1974). Icarus 22, 436.

Hohlfeld, R.G. & Terzian, Y. (1977). Icarus 30, 598.

Huang, S.-S. (1957). Pub. Astro. Soc. Pacific 69, 427.

Huang, S.-S. (1965). Astrophysical Journal 141, 985.

Huang, S.-S. (1977). IAU Colloquium No. 33, Rev. Mex. de Astr. y. Astrof., 3, special issue, p. 175.
Kuiper, G.P. (1935). Pub. Astro. Soc. Pacific 47, 15 and 121.
Norman, M.L., Wilson, J.R. & Barton, R.T. (1980), Astrophysical Journal 239, 968.
Rydgren, A.E., Strom, S.E. & Strom, K.M. (1976). Astrophysical Journal Supplement 30, 307.
Serkowski, K. (1976). Icarus 27, 13.
Skumanich, A. (1972). Astrophysical Journal 171, 565.
Trimble, V. & Cheung, C. (1976). IAU Symposium No. 73 "Structure and Evolution of Close Binary Systems", (editors: P. Eggleton, S. Mitton & J. Whelan) 369. D. Reidel, Dordrecht, Holland.
van de Kamp, P. (1975). Astronomical Journal 80, 658.

DISCUSSION

Sebastian von Hoerner: A few remarks on the angular momentum problem. Von Weizsacker wrote on this about twenty-five years ago. Low angular momentum in a star has nothing to do with the presence or absence of planets. I think that this was clear even before we knew about the solar wind and all that. You have turbulent friction, then you might have a mass flow inward, but a flow outward of angular momentum. This is a very probable situation; and then you have this surrounding gas, and whether the gas makes planets or not is a different question. But as long as you have a spectrum of turbulence then the planets cannot take up all the angular momentum. Concerning the eccentricities: in my view this means that the two clouds out of which the binary stars formed had been definite objects before the turbulence had decayed.

Harrington: Yes, that's what I said.

Von Hoerner: But the planets formed after the decay of the turbulence, and that is why they have such regular orbits.

Harrington: Collapse calculations often show the formation of a spinning ring at the center of the protostellar cloud. If this ring at some point just quietly divides into a binary we would expect a circular orbit, not the highly eccentric orbits observed. To get eccentric orbits you can either form the stars during infall, as you suggest, or you can do something violent to the material in the spinning ring. For example, if the ring breaks into three fragments, the gravitational interactions could eject one component (which could perhaps become a single star without planets) and leave an eccentric binary behind. One argument against this mechanism for making both single stars and binaries is the low velocity dispersion seen within groups of T Tauri stars: the ejected star would probably have a velocity comparable to the orbital velocity of the binary. The only point I wanted to make with regard to eccentricities was that the collapse calculations do not yet explain this feature of binary stars.

Eric Jones: I don't expect the colonization process to procede at all, or at least only in a very minor way, unless a substantial fraction of humanity is already living in space. If that is in fact the case, then planets may not be very necessary in the process.

Harrington: I am only considering them necessary from the point of view of providing raw material for the sustenance and replication of colonies.

Gerald Feinberg: What happens to the material when planets don't form? That is, the stuff in between double stars — where does it go?

Harrington: Well, it can either go out or it can go in. If it doesn't form substantial bodies, and it's very small, it will be blown out by radiation pressure. If it's somewhat larger, then the Poynting-Robertson effect will cause it to spiral into the star. That is, something orbiting a star runs into more radiation on its front side than on its trailing side and experiences a drag force. You just can't have dust in permanent orbit around a star.

Robert Shapiro: It seems conceivable that space colonists living in a permanent colony might survive living in a dense galactic cloud by sweeping up the material both as a source of power and for heavy elements, and just continue to live in that way without traveling to distant stars at all.

Jones: I seem to remember hearing an estimate of how long it would take to scoop up a drink of ethyl alcohol out of the interstellar medium.

Ben Zuckerman: We tried to put that estimate in the Astrophysical Journal, and I think they wouldn't let us do that. But they did let us put in an estimate of the total amount of ethyl alcohol in the cloud. Anyway, here's an example. If you took a scoop with the cross-sectional area of a football field, and hung it out behind your rocket ship and went through a few light-years of one of those dense clouds, dragging the scoop behind you, and then collected all the ethyl alcohol in the scoop, you would have less than one shot of booze.

17

Atmospheric Evolution, the Drake Equation, and DNA: Sparse Life in an Infinite Universe

Michael H. Hart

ABSTRACT

Computer simulations were made of the evolution of a planetary atmosphere, using as free parameters (a) the mass of the planet; (b) the semimajor axis of its orbit; and (c) the mass of the central star. The results indicate that only for a very small range of those parameters will the surface temperature of the planet remain continuously, for billions of years, in a range which permits H_2O to exist in the liquid phase.

Using a "Drake equation" approach, it is calculated that $N \sim 5 \times 10^5 \times f_{life}$, where N is the number of technological civilizations independently arising in a galaxy and f_{life} is the probability that life will spontaneously arise on a given habitable planet. However, the unlikelihood, even under ideal circumstances, of prebiotic processes lining up 600 or more nucleotides in the right order suggests that $f_{life} << 10^{-30}$. Hence N/galaxy is<<1. This agrees with and explains the common observation that the Earth has not been colonized by extraterrestrials.

The astronomical evidence favors an open (and apparently infinite) universe. There are probably, therefore, an infinite number of advanced extraterrestrial civilizations; but all are extremely far from us.

I. ATMOSPHERIC EVOLUTION

During the four and a half billion years of geologic time, the atmosphere of the Earth has changed markedly. Of the many processes which have played a role in that evolution, the most important are listed in Table 17.1.

Table 17.1. Important Processes in
Atmospheric Evolution

1.   Degassing of volatiles from interior

2.   Condensation of water vapor into oceans

3.   Solution of CO_2 and NH_3 in seawater

4.   Fixing of CO_2 in carbonate minerals (Urey reaction)

5.   Photodissociation of water vapor

6.   Escape of hydrogen into space

7.   Development of life and variations in the biomass

8.   Net photosynthesis and burial of organic sediments

9.   Chemical reactions between atmospheric gases

10. Oxidation of surface minerals

11. Changes in solar luminosity

12. Changes in albedo (cloud cover, ice cover, etc.)

13. The greenhouse effect

There are good reasons (Brown, 1952) to believe that at some early stage the Earth had almost no atmosphere, having lost whatever gaseous envelope, if any, that it started with. Our present atmosphere is derived from materials degassed from the interior of the Earth; perhaps largely from volcanoes, but also from fumaroles, and by a slow seepage through the crust.

Among the gases released was water vapor, most of which eventually condensed to form the oceans. The history of the other gases is highly complex. Some molecules were broken apart in the upper atmosphere by the action of sunlight. The lightest gases (hydrogen, helium) were able to escape into outer space (Spitzer, 1952). Other gases, such as carbon dioxide, are highly soluble in seawater and were able to react chemically with minerals dissolved in the oceans (Urey, 1951, 1952).

The origin of life, and subsequent biochemical processes (such as photosynthesis) have powerfully affected the composition of the atmosphere, as have various inorganic reactions such as the oxidation of surface minerals. Meanwhile, there have been marked changes in the Earth's surface temperature, caused in part by variations in the sun's luminosity, by variations in the Earth's reflectivity or albedo, and by the greenhouse effect.

In principle, if one knew the exact rate at which all these processes occurred, one could trace on a high-speed computer the entire evolution of the Earth's atmosphere over the past four and a half billion years. In an earlier paper (Hart, 1978), I have described in some detail the formulas which were used and the approximations which were made in constructing such a computer simulation. In view of the various approximations and uncertainties involved, one cannot expect the results of such a computer simulation to be reliable in every detail; but they are consistent with the available observational data, and they probably indicate the general pattern of our atmosphere's evolution fairly well.

According to the computer simulation, the Earth was probably a good deal warmer during the first 2.5 billion years of geologic time than it is today. It cooled down to about its present temperature roughly 2 billion years ago, when free oxygen first appeared in the atmosphere, and when various other gases capable of causing a large greenhouse effect were largely eliminated by oxidation.

Since the early Earth seems to have been quite warm, it is natural to wonder how much hotter it might have been if the Earth were situated somewhat closer to the Sun. It is fairly easy to modify the computer program so as to simulate the effect of a smaller Earth-Sun distance. The results are quite striking: If the Earth's orbit were only 5 percent smaller than it actually is, during the early stages of Earth's history there would have been a "runaway greenhouse effect", and temperatures would have gone up until the oceans boiled away entirely!

This result was not entirely unexpected. A similar conclusion (although based on a less detailed model) had been reached previously by Rasool and de Bergh (1970); and it is widely believed that a runaway greenhouse effect actually occurred on Venus, which is 28 percent closer to the Sun than the Earth is. More surprising, perhaps, were the results of computer runs which simulated the effect of a somewhat larger Earth-Sun distance. Those runs indicate that if the Earth-Sun distance were as little as one percent larger, there would have been runaway glaciation on Earth about 2 billion years ago. The Earth's oceans would have frozen over entirely, and would have remained so ever since, with a mean global temperature of less than -50°F. (Similar conclusions, although derived from quite different models, were reached earlier by Budyko (1969) and by Sellers (1969).) Taken together, these computer runs indicate that the habitable zone about our sun is not wide, as Huang (1959, 1960) had suggested, but is instead quite narrow.

What about the habitable zones about other stars? How large are they? Here too, it is possible to modify the original computer program to simulate the effect of the more intense radiation from a larger star, or the weaker radiation from a smaller one. The modifications needed are a bit tricky, since large stars evolve more rapidly than small stars, and their relative luminosities change with time. But the required modifications can be made, and again the results are quite striking. The computer simulations (Hart, 1979) indicate that a star whose mass, M_{star}, is less than $0.83\ M_{sun}$ will have no zone about it which is continuously habitable. If a planet is far enough from such a star to avoid a runaway greenhouse effect in its early years, it will inevitably undergo runaway glaciation somewhat later in its history. Nor, according to the computer results, does a star heavier than 1.2 solar masses have any continuously habitable zone about it.

Similar calculations indicate that the size of the planet itself has a profound effect on the evolution of its atmosphere. Unless a planet has a mass within the range $0.85\,M_{earth} < M_{planet} < 1.33\,M_{earth}$ it cannot — regardless of its distance from the Sun — maintain moderate surface conditions for more than 2 billion years.

II. CALCULATION OF N, USING A MODIFIED DRAKE EQUATION

The galaxy we are in, the Milky Way Galaxy, contains upwards of 100 billion stars, many of which appear to be quite similar to our Sun, and many of which may have planets orbiting about them. Within range of our large telescopes there are at least 10^9 other galaxies — possibly 10^{11} or more — together totaling at least 10^{20} stars.

In view of this enormous number of stars, it is quite natural to ask two questions: Of this vast multitude of stars, how many have planets near them which are suitable for the evolution of life? And on how many of those planets has life actually arisen? More intriguing still — since we are naturally more interested in <u>intelligent</u> life — are the questions: How many advanced civilizations can we expect to exist in the Milky Way Galaxy? How many can we expect in the entire universe?

To estimate N, the expected number of advanced civilizations in a typical galaxy the size of the Milky Way, many writers use as a starting point some version or modification of the well-known Drake equation. The version which I shall use is:

$$N = N_{gal} \cdot f_{popl} \cdot f_{PMR} \cdot f_{PS} \cdot f_{HP} \cdot f_{life} \cdot f_{intel} \cdot f_{tech} \tag{1}$$

In this equation, N_{gal} denotes the total number of stars in the galaxy; f_{popl} represents the fraction of those which are population type I stars; f_{PMR} denotes the fraction of population type I stars which are within the "proper" mass range (i.e., stars which are neither too large nor too small to have continuously habitable zones about them); f_{PS} represents the fraction of such stars which have planetary systems; f_{HP} denotes the fraction of planetary systems which include a habitable planet (i.e., a planet whose size, composition, and distance from its sun make it suitable for the development of life); f_{life} represents the fraction of habitable planets upon which life actually arises; f_{intel} is the fraction of those planets on which intelligent life forms (i.e., $\gtrsim$ human intelligence) eventually evolve; and f_{tech} denotes the fraction of those which develop and sustain advanced technologies (avoiding destruction by nuclear war, plagues, ecological disasters, etc.)

N_{gal} is usually estimated to be about 2×10^{11}. Of those, about 50 percent should probably be excluded because the gases from which they condensed had too low an abundance of heavy elements for large, solid planets like the Earth to be formed. According to the calculations described in section I, the "proper" mass range is $0.83\,M_{sun} < M_{star} < 1.2$ M_{sun}. Direct star counts in the solar neighborhood indicate that about 10 percent of stars fall within that range.

The value of f_p is still in doubt; suppose we estimate it to be about 10 percent. (As 50 percent or more of stars are members of double or multiple star systems, that can hardly be much of an underestimate. It might, though, be a serious overestimate — after all, as yet there is no reliable observation of a single planet outside our own solar system.)

If the fairly involved calculations described in section I are correct, only about one planetary system in a hundred (even if the central star is in the proper mass range) contains a habitable planet. That would make $f_{HP} \sim 10^{-2}$. The value of f_{life} is extremely speculative; for the moment, let us defer trying to estimate it. However, if life ever does arise on a planet, the process of Darwinian evolution should frequently lead to advanced life forms, and f_{intel} may well be as high as 10 percent. For the final factor, f_{tech}, a guess of 50 percent might be in order.

Since the value of f_{life} is so speculative, we might combine all the other factors in equation (1) together, and write it as

$$N = N_{combo} \cdot f_{life}. \qquad (2)$$

If we combine the various numerical estimates given above, we obtain the result: $N_{combo} \sim 5 \times 10^5$. However, as all the factors which go into N_{combo} are highly uncertain, its true value could be very different. Some optimists have estimated N_{combo} to be as high as 10^9, or perhaps even a bit larger; while if very conservative estimates are used for the various factors, N_{combo} could be only 10^1, or even less.

Now if f_{life} has a very low value — for example, 10^{-15} — this uncertainty in N_{combo} is unimportant. For in that case, any plausible value of N_{combo} results in $N \ll 1$. However, if f_{life} has a moderate value — say 10^{-2} — then the uncertainty in N_{combo} renders the "Drake equation approach" virtually useless as a method of deciding whether advanced civilizations are frequent in a typical galaxy, or whether the majority of galaxies do not contain even a single civilization. Nor, given the highly speculative nature of factors such as f_{intel} or f_{tech}, can we expect to obtain a reliable estimate of N_{combo} within the foreseeable future. What method, then, could we use to estimate the value of N?

III. OUR FAILURE TO OBSERVE EXTRATERRESTRIALS

I would suggest that in that case the best way to approach the problem of estimating N would not be by examining the factors which <u>cause</u> N to have a certain value, but rather by taking an empirical approach and considering the various <u>effects</u> which we might expect to observe if N had a given value.

If, for example, there were 100,000 advanced civilizations scattered about the Milky Way Galaxy, what observable effects might we expect to see? Well, if there were really so many technologically advanced races in our galaxy, then surely at least one of them would have explored and colonized the galaxy, just as we humans have explored and colonized this planet. Various estimates (Hart, 1975; Jones, 1976; Papagiannis, 1978) indicate that no more than a few million years would be needed to colonize most of the galaxy. Since that is very much less than the age of our galaxy ($\gtrsim 10^{10}$ years), if N were really as large as 100,000 then the solar system would have been colonized by extra-terrestrials a long time ago, and we would see them here today.

But, of course, we do not see any extraterrestrials, either on Earth or anywhere else in the solar system. There is no indication that the solar system was ever visited by extraterrestrials; and, quite obviously, we have not been colonized. It can reasonably be concluded, therefore, that N is not equal to 100,000. The same argument, of course, would rule out any other large number. It would not, though, completely rule out the possibility that there were a small number of civilizations in our galaxy, none of which were interested in interstellar exploration and colonization (nor ever had been, in all the ages since they first acquired the technological capability).

N, therefore, is a small number, possibly a very small number; and our conclusion, since it has an empirical basis (i.e., the absence of extraterrestrials on Earth) cannot be upset by any unreliable calcula-tions based on the Drake equation. Nevertheless, it would certainly be interesting to know just how low N is. I would like to suggest that a realistic calculation of f_{life} indicates that it is an extremely low number, and that N therefore is also extremely low.

IV. CALCULATION OF f_{life}

Before attempting to compute f_{life}, we should perhaps first explain what we mean by the word "life." It is difficult to give an exact definition of this term (see Feinberg and Shapiro (1980) for an in-teresting and novel approach), but we might roughly say that a living organism is an object which feeds and reproduces. (An object "feeds" if it ingests and chemically transforms material in its environment into chemicals which it is itself composed of).

The living organisms which we see on Earth are all composed of complex carbon compounds in an aqueous medium. A wide variety of such compounds are found in most organisms, but the two most significant types are: (a) the proteins, which are large, complex molecules consisting of long strings of simpler components called amino acids; and (b) the nucleic acids, which consist of long strings of simpler components called nucleotides. (The most important type of nucleic acid, DNA, contains 4 different nucleotides, each occurring many times in a single molecule of DNA). The proteins perform a crucial role in catalysing essential biochemical reactions, while the nucleic acids perform an even more vital role by storing the hereditary information which allows organisms to reproduce, and by directing the synthesis of proteins. Nucleic acids are the primary genetic material, and they contain (in coded form) instructions for synthesizing the organism and its components. The code is based on the number of each of the 4 types of nucleotides in a given strand, and on the _order_ in which those different nucleotides are arranged.

Now if there is life on other planets in the universe, it is perfectly possible that the organisms on such planets use quite different compounds to perform the tasks which in terrestrial organisms are carried out by the proteins and the nucleic acids. But since those tasks are so difficult, detailed and varied, the compounds carrying them out would of necessity have to be just about as large and as complex in structure as are the proteins and nucleic acid molecules which we find in terrestrial organisms.

How large, then, is f_{life}, which is defined as the probability that life will actually arise on a given planet which has a wholly suitable environment. We may safely assume that on such a planet the surface temperatures are suitably moderate, that liquid water is present in ample quantity, and that simple compounds of carbon, oxygen, hydrogen and nitrogen are abundant. Many experiments (see Miller and Orgel, 1974 for a partial list) have shown that a combination of such chemicals will, in the presence of electric discharges, react to produce a variety of more complex organic molecules, including amino acids. Under suitable conditions, short chains of amino acids have also been produced.

This is an encouraging start. However, in order to have living organisms, some sort of genetic material – such as DNA – must be present also. Experiments simulating primitive Earth conditions have not, to date, resulted in the formation of nucleotides; but simpler compounds related to them have been produced in such experiments, and it is not unduly optimistic to assume that nucleotide molecules too will naturally be formed on a suitable planet.

To induce those molecules to polymerize into nucleic acid strands (under the assumed primitive Earth conditions) is a bit of a problem, but not a hopeless one. It is, though, crucial for the proper functioning of the resulting nucleic acid molecule that the various nucleotides in the strand are arranged in the correct order. Two different nucleic acid strands, even if of exactly the same length, will not normally be biologically equivalent unless they contain the same nucleotide residues arranged in the same order.

The great majority of possible nucleic acid molecules are quite useless (or even harmful) biologically. Most of the others are useful only in an organism which already has many other genes. Let us suppose, however, that there exists a particular DNA molecule – "genesis DNA" – which if introduced into some primitive conglomeration of proteins, lipids, nucleotides, and their building blocks will both replicate properly and perform some useful biological function. In other words, we are supposing that the formation of a single molecule of genesis DNA and its introduction into a suitable environment will suffice to create a viable organism and to get the process of Darwinian evolution started.

To simplify our calculations, let me make a few more assumptions (admittedly, rather optimistic ones): (a) Under the conditions prevailing on a primitive Earth-like planet, not just amino acids but also nucleotides will be readily formed. (b) Those same conditions will favor the polymerization of nucleotides. (c) Uniform helicity of the resulting strands is thermodynamically favored. (d) A strand of genesis DNA is quite short, as genes go, containing only 600 nucleotide residues. (e) There exists some chemical effect which favors the formation of nucleic acid strands of that length. If these assumptions are valid, then a large number of strands of nucleic acid, each consisting of about 600 residues, will be formed spontaneously on any suitable planet. Let us calculate the probability that one of those strands will have its residues arranged in the right order; that is, in the same order as in genesis DNA.

There are only about 2×10^{44} nitrogen atoms near the surface of the Earth, or in its atmosphere. As a single 600-residue strand of nucleic acid includes more than 2000 nitrogen atoms, there could have been no more than about 10^{41} strands of DNA existing together on the primitive Earth at any given moment. If every such strand could split up and recombine with other fragments at a rate of 30 times a second, then in one year (roughly 3×10^{7} seconds) a maximum of 10^{50} different strands could be formed, and in 10 billion years a maximum of 10^{60}. (This, obviously, is a strong maximum).

The number of conceivable arrangements of the four different nucleotides into a strand of DNA 600 residues long is 4^{600}, which is about 10^{360}. The chance that a <u>particular</u> one of them would be formed spontaneously — even in 10 billion years — was therefore extremely small, 10^{-300}. However, the chance of forming genesis DNA is not necessarily that low. It has been demonstrated that at some positions in a strand of nucleic acid it is possible to replace one nucleotide residue by another without changing the biological effect of the strand. Let us suppose (very optimistically) that in a strand of genesis DNA there are no fewer than 400 positions where any one of the four nucleotide residues will do, and at each of 100 other positions either of two different nucleotides will be equally effective, leaving only 100 positions which must be filled by exactly the right nucleotides. This appears to be an unrealistically optimistic set of assumptions; but even so, the probability that an arbitrarily chosen strand of nucleic acid could function as genesis DNA is only one in 10^{90}. Even in 10 billion years, the chance of forming such a strand spontaneously would be only $10^{-90} \times 10^{60}$, or 10^{-30}.

There are several reasons why the true value of f_{life} is very much lower than 10^{-30}. In the first place, we have ignored all the difficulties involved in producing nucleotides abiotically, in concentrating them in a small region, of preventing their spontaneous destruction, and in getting them to polymerize in an aqueous environment.

In the second place, a DNA molecule cannot direct protein synthesis unless certain other complex organic molecules called transfer RNA are present; nor can it even replicate itself spontaneously in the absence of certain other organic catalysts (DNA polymerases). Unless these other compounds had already been formed (how?) and were in the immediate vicinity, even if a molecule of genesis DNA happened to be formed it would be unable to function. And in the third place, the assumption that there exists a gene — genesis DNA — which, without any other genes present, can produce a viable organism is highly optimistic. The simplest known organism which is capable of independent existence includes about 100 different genes. For each of 100 different specific genes to be formed spontaneously (in ten billion years) the probability is $(10^{-30})^{100} = 10^{-3000}$. For them to be formed at the same time, and in close proximity, the probability is very much lower.

V. PROBABILITY AND SELECTION

The conclusion reached above, that the probability of life arising on a given planet — no matter how favorable conditions on that planet might be — is less than one in 10^{30}, is perhaps somewhat surprising. If f_{life} is so low, you may ask, what are we doing here?

This leads to an interesting philosophical question: If we calculate the probability of an event occurring to be a very low number, and the event then occurs, does it show that our calculation is wrong? For example, a simple calculation shows that the probability of tossing an honest coin 40 times and getting 40 consecutive heads is $(\frac{1}{2})^{40}$, or about one in a trillion. Suppose, though, that you took a particular coin, flipped it 40 times, and got 40 heads. Would you then rush about excitedly, telling everyone about the "almost unbelievable" coincidence which had occurred, and send in a report to a scientific journal? Of course not! You would simply conclude that the coin was not balanced, and that your calculations therefore did not apply.

Suppose, however, that you made not just one set of 40 flips, but 10^{12} different runs, each of 40 flips. And further suppose that one of those runs resulted in 40 consecutive heads. In that case you would conclude that the coin was honest, that your calculations were correct, and that no unbelievable coincidence had occurred.

Similarly, if we were shown an (undoctored) film displaying a run of 40 consecutively heads, we would normally interpret it not as evidence of a remarkable coincidence, but merely as evidence that the coin was unbalanced. However, if we knew that the maker of the film had made and photographed 10^{12} runs, each of 40 flips, but only let us see the film of the one successful run, we would see no reason to doubt the correctness of our calculations.

VI. LIFE IN THE INFINITE UNIVERSE

Why do I suggest such a fanciful possibility? Because the universe we live in is not finite, but infinite! Modern astronomical observations strongly support the so-called "big-bang" cosmology, and the majority of the evidence indicates that our universe is open and will continue to expand indefinitely (Gott et al., 1974). Analysis shows that, unless a very unusual topology is assumed, such an open universe must be infinite in extent, with an infinite number of galaxies, an infinite number of stars, and an infinite number of planets. In an infinite universe, any event which has a finite probability – no matter how small – of occurring on a single given planet must inevitably occur on some planet. In fact, such an event must occur on an infinite number of planets. (See Ellis and Brundrit (1979) for an interesting discussion of this point).

We are therefore in the position of the hypothetical film-viewer described above. There are an infinite number of habitable planets in the universe. On each of these, nature patiently tosses her tetrahedral dice for ten billion years, trying to line up 600 nucleotides in the proper order to make genesis DNA. In the great majority of cases the attempt is unsuccessful; but these "runs", of course, are never seen. Only in that rare case when a run is successful, and life does get started on a planet, is there anyone around to view the film.

The universe, therefore, contains an infinite number of inhabited planets, but the chance that any specific galaxy will contain life is extremely small. Most intelligent races should see no other civilizations in their galaxy; indeed, they should see no others in the entire portion of the universe (including perhaps 10^{22} stars) which they are able to observe with their telescopes. This theoretical prediction is, of course, in complete agreement with our failure to observe extraterrestrials, and with all our other observational evidence.

CONCLUSION

All of the calculations made above are based on existing theories. No extraordinary assumptions have been made, nor have any unknown effects or processes been postulated. Normally, when theoretical conclusions based on existing theories are in complete accord with the observations the conclusions are readily accepted.

Why, then, are so many people reluctant to believe that N is a low number? I would suggest that this reluctance is primarily a result of wishful thinking: a galaxy teeming with bizarre life forms sounds a lot more interesting than one in which we are alone. But N can be a high number only if $f_{life} >> 10^{-30}$, and that can only be the case if there exists some abiotic process – as yet totally unknown – which lines up nucleotides in a sequence which is biologically useful. Although we cannot absolutely prove that no such process exists, we should certainly be reluctant to postulate an unknown process when all the observed facts can be explained without it.

REFERENCES

Brown, H. (1952). Rare gases and the formation of the Earth's atmosphere. In The Atmospheres of the Earth and Planets, (editor: G.P. Kuiper), 2nd ed., 258-266. University of Chicago Press, Chicago.

Budyko, M.I. (1969). The effect of solar radiation variations on the climate of the Earth. Tellus 21, 611-619.

Ellis, G.F.R. and Brundrit, G.B. (1979). Life in the infinite universe. Quarterly J. Royal Astronomical Soc. 20, 37-40.

Feinberg, G. and Shapiro, R. (1980). Life Beyond Earth. William Morrow and Co., New York, chapter 6.

Gott, J.R., Gunn, J.E., Schramm, D.M. and Tinsley, B.M. (1974). An unbound universe? Astrophysical Journal 194, 543-553.

Hart, M.H. (1975). An explanation for the absence of extraterrestrials on Earth. Quarterly J. Royal Astronomical Soc. 16, 128-135. (Reprinted in this volume.)

Hart, M.H. (1978). The evolution of the atmosphere of the Earth. Icarus 33, 23-29.

Hart, M.H. (1979). Habitable zones about main sequence stars. Icarus 37, 351-357.

Huang, S.-S. (1959). Occurrence of life in the universe. American Scientist 47, 397-402.

Huang, S.-S. (1960). Life outside the solar system. Sci. Amer. 202, 55-63.

Jones, E.M. (1976). Colonization of the galaxy. Icarus 28, 421-422.

Miller, S.L. and Orgel, L.E. (1974). The Origins of Life on the Earth, Prentice-Hall, Englewood Cliffs, N.J., 100-102.

Papagiannis, M.D. (1978). Could we be the only advanced technological civilization in our galaxy? In Origin of Life (editor: H. Noda), Center Acad. Publ., Tokyo, Japan.

Rasool, S.I. and de Bergh, C. (1970). The runaway greenhouse and the accumulation of CO_2 in the Venus atmosphere. Nature 226, 1037-1039.

Sellers, W.D. (1969). A global climate model based on the energy balance of the Earth-atmosphere system. J. Applied Meteorology 8, 392-400.

Spitzer, L. (1952). The terrestrial atmosphere above 300 km. In The Atmospheres of the Earth and Planets (editor: G.P. Kuiper), 211-247. University of Chicago Press, Chicago.

Urey, H.C. (1951). The origin and development of the Earth and other terrestrial planets. Geochim. Cosmochim. Acta 1, 209-277.

Urey, H.C. (1952). On the early chemical history of the Earth and the origin of life. Proc. Nat. Acad. Sci. U.S.A. 38, 351-363.

DISCUSSION

Unidentified speaker: Is there anything particularly wrong with letting a planet freeze for a billion years and then thawing it out?

Hart: There are some spores that can survive freezing for a very long time. Once a planet gets frozen it's in a very stable equilibrium, and even a considerable increase in the stellar luminosity won't get it out of equilibrium because the albedo of ice is so high, about 70 percent. It went into runaway glaciation when it was a medium-albedo planet, and now its a high-albedo planet; so there has to be a big increase in luminosity to compensate. Nonetheless, all main sequence stars eventually, if you are willing to wait a long period of time, are going to have that large increase in luminosity. There may be some of those frozen planets around which may thaw out in a few billion years. I suspect that when they do thaw out, with a much hotter sun, once the ice caps melt the albedo goes down and the temperature gets still higher. So after the icecaps melt you will probably get a runaway greenhouse effect because the temperature is so high.

Unidentified speaker: Why can't you heat up the planet by having volcanic gas come out from the interior and increase the greenhouse effect, rather than by increasing the luminosity of the star?

Hart: That is possible. I haven't looked into that matter in detail. At present a greenhouse effect on the Earth with only CO_2 isn't going to get very far at all. Water vapor is the main cause of the Earth's greenhouse effect, and if that's frozen out you need an awful lot of CO_2 to raise the temperature appreciably. The temperature of a planet suffering from runaway glaciation is down around 200° Kelvin, so you need a monster greenhouse effect to get it out.

Eric Jones: How long does runaway glaciation take?

Hart: There have been no accurate calculations on just how long it takes. My simulations and top of the head feelings indicate that once the runaway glaciation starts it goes pretty much to completion in about 50,000 years or less, maybe significantly less.

Edward Argyle: Will the oceans freeze to the bottom?

Hart: I don't know. Given time I suppose that they freeze to the bottom. Perhaps the continents will only have a thin veneer of frost on them. It doesn't much affect the overall solutions.

Some of the participants at the symposium, College Park, Md., November 2-3, 1979.

List of Participants at the Symposium

Where Are They?...

Edward Argyle
Dominion Radio Astrophysical
    Observatory
Penticton, B.C., Canada V2A 6K3

Ronald Bracewell
Radio Astronomy Institute
Stanford, CA 94305

Freeman Dyson
Institute for Advanced Study
Princeton, NJ 08540

Gerald Feinberg
Department of Physics
Columbia University
New York, NY 10027

J. Patrick Harrington
Astronomy Program
University of Maryland
College Park, MD 20742

Eric M. Jones
Los Alamos Scientific
    Laboratory
Los Alamos, NM 87545

Siegfried Bauer
NASA Goddard Space Flight
    Center
Greenbelt, MD 20771

John Carlson
Astronomy Program
University of Maryland
College Park, MD 20742

David Eichler
Astronomy Program
University of Maryland
College Park, MD 20742

J. Richard Gott, III
Department of Astrophysical
    Sciences
Princeton University
Princeton, NJ 08540

Michael H. Hart
Department of Physics
Trinity University
San Antonio, TX 78284

Shiv Kumar
Department of Astronomy
University of Virginia
Charlottesville, VA 22903

James Oberg
NASA Johnson Space Center
Houston, TX 77058

Michael D. Papagiannis
Department of Astronomy
Boston University
Boston, MA 02215

Robert Rood
Department of Astronomy
University of Virginia
Charlottesville, VA 22903

Robert Shapiro
Department of Chemistry
New York University
New York, NY 10003

Cliff Singer
Princeton University
Plasma Physics Laboratory
P.O. Box 451
Princeton, NJ 08544

Jill Tarter
NASA Ames Research Center
MS TR-002
Moffett Field, CA 94035

Virginia Trimble
Astronomy Program
University of Maryland
College Park, MD 20742

Allan Walstad
Physics Department
University of Pittsburgh at
   Johnstown
Johnstown, PA 15904

Ben Zuckerman
Astronomy Program
University of Maryland
College Park, MD 20742

Patrick Palmer
Astronomy & Astrophysics
   Department
University of Chicago
Chicago, IL 60637

Cyril Ponnamperuma
Laboratory of Chemical
   Evolution
University of Maryland
College Park, MD 20742

David Schwartzman
Department of Geology and
   Geography
Howard University
Washington, DC 20001

Robert Sheaffer
1341 Poe Lane
San Jose, CA 95130

Harlan Smith
Astronomy Department
University of Texas
Austin, TX 78712

James Trefil
University of Virginia
Charlottesville, VA 22903

Sebastian von Hoerner
National Radio Astronomy
   Observatory
Green Bank, WV 24944

Hubert Yockey
Army Pulse Radiation Facility
Aberdeen Proving Ground,
   MD 21005

Index

About the Editors and Contributors

MICHAEL H. HART did his undergraduate work at Cornell University and later obtained a Ph.D. in astronomy from Princeton. He also has graduate degrees in physics and in law. He has written numerous articles on astronomy for professional journals and is also the author of a book on history: The 100: A Ranking of the Most Influential Persons in History. Dr. Hart has worked at Hale Observatories in California, at the National Center for Atmospheric Research in Colorado, and at NASA's Goddard Space Flight Center in Maryland. At present Dr. Hart teaches physics and astronomy at Trinity University in San Antonio, Texas. He is married and has two children.

BENJAMIN ZUCKERMAN is a professor in the Physics and Astronomy Department at the University of Maryland. When not ruminating about intelligent life in the Universe, his major scientific interests are the birth and death of stars. He has been a co-discoverer of various molecules in interstellar space including formaldehyde, ethyl alcohol, and formic acid. Dr. Zuckerman has received fellowships from the Alfred P. Sloan and John S. Guggenheim Foundations and various awards such as the Helen B. Warner Prize of the American Astronomical Society. He enjoys hiking in remote areas of the Grand Canyon.

EDWARD ARGYLE has a master's degree in nuclear physics and a Ph.D. in astronomy. He is a radio astronomer, and has spent most of his career at the Dominion Radio Astrophysical Observatory in Penticton, British Columbia. He was a senior research officer there until his recent retirement.

179

RONALD N. BRACEWELL is a professor of electrical engineering at Stanford University. A native of Australia, he has degrees in mathematics, physics and electrical engineering. Dr. Bracewell is the author of The Galactic Club: Intelligent Life in Outer Space, and has also written books on radio astronomy and applied mathematics.

FREEMAN DYSON has for many years been a professor of physics at the Institute for Advanced Study in Princeton, N.J. He has also been a consultant to several government agencies, including the Arms Control and Disarmament Agency, the Defense Department, and NASA. Professor Dyson is the author of many articles, as well as a book, Disturbing the Universe. He has been awarded several distinguished prizes including the Max Planck medal of the German Physical Society, the Robert Oppenheimer Memorial Prize, and the Hughes Medal of the Royal Society of London.

GERALD FEINBERG is chairman of the department of physics at Columbia University. He has written numerous articles for scientific journals, as well as three books for the general reader. Two of those books, The Prometheus Project and Consequences of Growth, deal with the future, its possibilities, and some of the resulting ethical problems. He is also co-author, with Robert Shapiro, of Life Beyond Earth: The Intelligent Earthling's Guide to Life in the Universe.

J. RICHARD GOTT, III is an associate professor at Princeton University in the Department of Astrophysical Sciences. Before coming to Princeton he did research at the California Institute of Technology and at Cambridge University in England. Professor Gott's research specialty is cosmology, but he has also written articles on many other branches of astronomy and astrophysics. In 1975 he received the Trumpler award of the Astronomical Society of the Pacific.

PAT HARRINGTON is associate professor of astronomy at the University of Maryland. A native of Salem, Ohio, he followed his undergraduate degree in physics from the University of Chicago with graduate work in astronomy at Ohio State University. Upon receipt of his doctorate in 1967, he joined the Maryland faculty where he has remained except for a year at University College London. He has authored numerous papers on the astrophysics of planetary nebulae and on the transfer of radiation in stellar atmospheres. He also admits to an interest in late classical antiquity and in the paintings of Claude Lorrain. He never jogs.

ERIC M. JONES did his undergraduate work at the California Institute of Technology and then obtained a Ph.D. in astrophysics from the University of Wisconsin. Since then he has been a research scientist at the Los Alamos National Laboratory in New Mexico. For five years he was the group leader for the underground nuclear test containment program. Currently he specializes in numerical simulation of the effects of nuclear explosions.

JAMES OBERG works for McDonnell-Douglas as a contractor to NASA in mission control, specializing in computer control of orbital maneuvering rockets and attitude control rockets. He writes a monthly column for OMNI magazine, and has contributed articles to New Scientist, Star and Sky, Space World, The Skeptical Inquirer, Astronomy, and many other magazines. His two most recent books are: New Earths, which discusses the possibilities of planetary engineering, and Red Star in Orbit, a survey for the layman of the Soviet manned space program.

MICHAEL D. PAPAGIANNIS was born and raised in Greece, but came to the United States as a young man and received a Ph.D. in physics and astronomy from Harvard. He is now chairman of the astronomy department at Boston University. Dr. Papagiannis' research specialties are space physics and radio astronomy. In addition to his many journal articles, he is the author of a book: Space Physics and Space Astronomy. He is also the editor of two other books, one on relativistic astrophysics and one entitled Strategies for the Search for Life in the Universe.

CYRIL PONNAMPERUMA is Professor of Chemistry and Director of the Laboratory of Chemical Evolution at the University of Maryland. His main research interests are in the study of the origin of life and the possibility of life beyond the earth. He has been a principal investigator in the Apollo Program and has been associated with many of NASA's planetary missions. In 1980 he received the first A.I. Oparin award presented by the International Society for the Study of the Origin of Life for "the best sustained research program" on the origin of life.

ROBERT SHAPIRO is a professor of Chemistry at New York University where he does research concerning the chemistry of our hereditary material and the nature of environmental pollutants. He is co-author, with Gerald Feinberg, of Life Beyond Earth: The Intelligent Earthling's Guide to Life in the Universe.

ROBERT SHEAFFER, a leading skeptical investigator of UFOs, is a computer software specialist and a freelance writer. A founding member of the UFO Subcommittee of the Committee for The Scientific Investigation of Claims of the Paranormal, he actively pursues a lifelong interest in astronomy and the question of life in other worlds. His book, The UFO Verdict, has just been published.

CLIFFORD E. SINGER received his Ph.D. in biochemistry at Berkeley in 1971 and then spent five years at Queen Mary College in London. He now works on controlled fusion at Princeton. He has published research on molecular biology, genetic evolution, solar and lunar physics, plasma physics, extraterrestrial resources, interstellar propulsion, and communication with extraterrestrial intelligence. His hobby is trying to minimize the probability of human extinction.

VIRGINIA TRIMBLE is a native Californian, having received degrees from UCLA (B.A. 1964), Caltech (M.S. 1965; Ph.D. 1968), and Cambridge University (M.A. 1969). She currently shares appointments at the University of California, Irvine, and the University of Maryland with her husband, physicist Joseph Weber. Her previous affiliations include Smith College and the Institute of Theoretical Astronomy (Cambridge). She has published research and review papers in the fields of cosmology, evolution of stars and galaxies, and life in the universe.

SEBASTIAN VON HOERNER is a scientist at the U.S. National Radio Astronomy Observatory. He was born in Germany and studied theoretical physics at Gottingen. He worked at the Max-Planck-Institut in Gottingen and the Astronomisches Recheninstitut in Heidelberg where he investigated diverse problems in turbulence, shock fronts, astrophysical hydrodynamics, stellar dynamics, and stellar evolution. Dr. von Hoerner's work at the NRAO has concerned radio astronomy, cosmology, life in space, and the design and improvement of radio telescopes. He has published two papers on musical scales and enjoys sailing and hiking.